RATIONAL THERMODYNAMICS

McGRAW-HILL SERIES IN

MODERN APPLIED MATHEMATICS

MENAHEM M. SCHIFFER Consulting editor

FLÜGGE-LOTZ Discontinuous and Optimal Control

LEITMANN An Introduction to Optimal Control

TRUESDELL Rational Thermodynamics

RATIONAL THERMODYNAMICS

A Course of Lectures
on Selected Topics

C. TRUESDELL

McGRAW-HILL BOOK COMPANY

New York St. Louis San Francisco
London Sydney Toronto Mexico Panama

Rational Thermodynamics

Library of Congress Catalog Card Number 79-77961

65300

1 2 3 4 5 6 7 8 9 0 M A M M 7 6 5 4 3 2 1 0 6 9

In the popular treatise, whatever shreds of the science are allowed to appear, are exhibited in an exceedingly diffuse and attenuated form, apparently with the hope that the mental faculties of the reader, though they would reject any stronger food, may insensibly become saturated with scientific phraseology, provided it is diluted with a sufficient quantity of more familiar language. In this way, by simple reading, the student may become possessed of the phrases of the science without having been put to the trouble of thinking a single thought about it. The loss implied in such an acquisition can be estimated only by those who have been compelled to unlearn a science that they might at length begin to learn it.

The technical treatises do less harm, for no one ever reads them except under compulsion. From the establishment of the general equations to the end of the book, every page is full of symbols with indices and suffixes, so that there is not a paragraph of plain English on which the eye may rest.

MAXWELL

A new scientific truth does not triumph by convincing its opponents and making them see the light, but rather because its opponents finally die, and a new generation grows up that is familiar with it.

PLANCK

PREFACE

These lectures are meant as a sequel to my *Six Lectures on Modern Natural Philosophy*, Springer-Verlag, New York, 1966, and especially as a replacement for Lecture III therein. They began as the first nine in the series of public lectures delivered in 1966–1967 at the Drexel Institute of Technology, Philadelphia, under the title "Topics in Modern Natural Philosophy." The part of that series devoted to electromagnetism I have withheld here as being unripe for the press; in return, I have completely rewritten Lectures 3 and 4, have recast Lecture 5, and have added a new lecture, here the sixth, all these containing material which did not exist last year. Most of the lectures in their present forms I have delivered in 1968 at one or more of the Universities of Rome, Pisa, and Uppsala, the Technical Universities of Lund and Stockholm, and the Nordita at Copenhagen.

I am indebted to Messrs. BOWEN, ERICKSEN, GURTIN, MÜLLER, and WANG for helpful criticism and to the U.S. National Science Foundation for its support of my work.

For the full background in mechanics, the reader may refer to an article by NOLL & me, *The Non-Linear Field Theories of Mechanics*, in FLÜGGE's *Handbuch der Physik*, volume III/3, Springer-Verlag, Berlin-Heidelberg-New York, 1965; for an introduction, to my *Elements of Continuum Mechanics*, Springer-Verlag, New York, 1966. For work done since the relevant sections of those expositions were written, I have adjoined references here, but I have not attempted to write a history of thermodynamics or even to mention the name of every person who has contributed to it recently.

Since I have held the mathematics to an elementary level so as to keep the physical side always in sharp relief, these lectures plainly do not make up a deductive treatise. Neither are they a mountain railway that carries the flaccid tourist to peak after peak at no cost of toil on his own part. Rather, they offer a survey from below, brought to life by a few sample climbs.

C. TRUESDELL
Amsterdam, 1 April 1968

CONTENTS

LECTURE 1

THERMODYNAMICS OF HOMOGENEOUS PROCESSES

It is often claimed that while mechanics is a mathematical theory, thermodynamics is a physical one. Although this distinction is confirmed by the usual treatments of the two subjects, it is historical only, in no way intrinsic to the regions of experience to which the two names refer. Every aspect of nature may be approached by poetry or experiment as well as by reason, and indeed such is the usual order in history. The moon was first conceived as the home of a goddess who controlled its motion as well as participating in other and less regular ways in human and superhuman affairs; later, the goddess was banished, replaced by great tables of precisely observed spherical co-ordinates, which baffled all attempts to reduce them to reason; and only after another thousand years were those piles of numbers shown, at last, to be only slightly inaccurate corollaries of the formulae $\mathbf{f} = m\mathbf{a}$, $f = m_1 m_2 G/r^2$, and $\mathbf{L} = (\mathbf{I}\omega)^{\cdot}$. The discovery of these formulae and the development of their consequences to the point of predicting the lunar motion destroyed the value neither of Diana nor of telescopic observation, but it did suggest that other aspects of that goddess were more important than her guidance of the moon, and it did leave the gazers into telescopes free to concentrate on properties of the moon more interesting than its spherical co-ordinates.

There is certainly a physical side to mechanics, which is not harmed by the existence of developed mathematical theories. Once, long ago, there was only a physical side. In the seventeenth and eighteenth centuries, mathematical theories of mechanics were created, studied, applied, and reformulated again and again. As a result, mechanics became more precise, briefer, easier to learn, and more widely applicable. While at first rational mechanics concerned mainly equilibrium and then the motion of very simple systems like single mass-points or rigid spheres or pendulums, it was extended and generalized so as finally to include deformable bodies in rapid and even discontinuous motion. It grew and spread to the point that a boy can be taught in one year at the university theories and methods which enable him to formulate and solve correctly problems before which such great minds as NEWTON's and HUYGENS' stood helpless; it is still growing today.

Thermodynamics has had a different history. It began out of steam tables, venous bleeding, and speculations about the universe and has always had a hard time striking a mean between these extremes. While its claims are often grandiose, its applications are usually trivial. The classical illustrations all concern systems which from the standpoint of mechanics are so special as to be degenerate, yet thermodynamicists are prone to claim that their science somehow implies mechanics as a corollary.

The first impetus toward a combined thermodynamics and mechanics of deforming bodies came with gas dynamics, a science first formulated, after major earlier work by such men as HUGONIOT, by HADAMARD in his lectures of 1898–1900, which grew in part from his association just before then with DUHEM. With the coming of flight near or beyond sonic speeds, this theory experienced a great revival in the 1940s, if more in breadth than in depth. A determinate system of general equations for gas dynamics requires a relation between the pressure and the density of the flowing, deforming gas. In a gas in equilibrium, such a relation depends upon the temperature as well. When he faced this problem over a hundred years earlier, EULER had been able to suggest nothing better than to regard the temperature as an assigned function of place. In this way he explained convection in a tube, but of course he could do nothing in more typical flows, where the temperature of a gas is determined by boundary and initial conditions and must be found as a function of place and time as a part of the solution of a problem rather than being prescribed. EULER did not know the general principle of energy; DUHEM and HADAMARD did; thus they were able to state instead the hypothesis—and they emphasized that it was no more than a hypothesis—that in a gas flow, the same equations relate the pressure p, the specific volume v, and the temperature θ as in equilibrium. This forced them to introduce the specific entropy η and the specific internal energy ε and to use the equations relating them according to the *thermostatics* of homogeneous fluids:

$$\varepsilon = \varepsilon(v, \eta), \qquad p = -\partial_v \varepsilon, \qquad \theta = \partial_\eta \varepsilon, \tag{1.1}$$

where the function $\varepsilon(\cdot)$ is to be determined from the results of static experiments. This decisive step has been questioned but rarely by specialists in gas dynamics, and the entire theory of bodies flying in the lower atmosphere rests upon it. It is often put forward nowadays as the claim that volume elements in an irreversible process suffer only reversible changes and hence may be regarded as thermodynamic systems in equilibrium, but of course such men as HADAMARD and DUHEM were not ashamed of a good guess and felt no need to hide it under a thermodynamic fog screen.

Recent work on large deformations of materials with memory, as well as the theory of very rarefied gases, suggests that the relations (1.1) are only approximate in some sense, but those relations as they stand do not make clear what is being approximated. For ten years or more, everyone studying materials has seen that both thermodynamics and mechanics must be brought to bear, but these sciences as hitherto taught do not mesh. Thus, if we take up almost any recent book with "continuum mechanics" or "material science" in its title, we find a chapter on thermodynamics, but that chapter presents a curious contrast with the same author's pages, earlier in the book, on pure mechanics. There the reader is faced by mappings, fields of vectors and tensors, Jacobians, differential invariants, perhaps even Christoffel symbols and affine connections; of course he is presumed familiar with calculus as calculus has been taught for the last fifty years. He can understand dynamical equations in tensorial form:

$$\operatorname{div} \mathbf{T} + \rho \mathbf{b} = \rho \ddot{\mathbf{x}} \qquad \text{or} \qquad T^{km}{}_{,m} + \rho b^k = \rho \ddot{x}^k. \tag{1.2}$$

He is informed, in each case, what the dependent and independent variables are, he is presented with explicit differential equations and boundary-value problems, he is shown many special solutions in concrete cases and is directed to grand tomes where he can find thousands more such solutions not given in the book he is reading, and often he is told about some major problems still unsolved and is challenged to solve them himself. The same reader of the same book then reaches the chapter on thermodynamics, where he is faced with the "axiom"

$$T\, dS \geqq \delta Q. \tag{1.3}$$

He is told that dS is a differential, but not of what variables S is a function; that δQ is a small quantity not generally a differential; he is expected to believe not only that one differential can be bigger than another, but even that a differential can be bigger than something which is not a differential. He is loaded with an arsenal of words like piston, boiler, condenser, heat bath, reservoir, ideal engine, perfect gas, quasi-static, cyclic, nearly in equilibrium, isolated, universe—words indeed familiar in everyday life, doubtless much more familiar than "tangent plane" and "gradient" and "tensor", which he learned to use accurately and fluently in the earlier chapters, but words that never find a place in the mathematical structure at all, words the poor student of science is expected to learn to hurl for the rest of his life in a rhetoric little sharper than that wielded by a housewife in the grocery store. The mathematical structure, in turn, is slight. There are no general equations to be solved, no boundary-value or initial-value problems set, no general theorems characterizing classes of solutions. The examples or exercises consist in no more than calculating partial derivatives or integrals of given functions or their inverses and plugging numbers into the results. The references cited lead to other books containing just the same material, perhaps otherwise explained and ordered, but no broader or clearer in concept, and equally unmathematical. No problems, in the sense that the word "problem" has in the theories of mechanics or electromagnetism or optics or heat conduction, are solved. Neither are any open problems stated. The reader must presume that thermodynamics is an exhausted as well as exhausting subject, with nothing left to be done.

As if to emphasize the difference between mechanics and thermodynamics, even the notations of calculus change from one to the other. Not only do differentials replace derivatives, but even derivatives look different, *e.g.*,

$$\frac{\delta^{\mathrm{rev}}Q}{dV} = T\left(\frac{\partial S}{\partial V}\right)_P. \tag{1.4}$$

The difference is not merely formal (note the three different *d*'s). It represents a difference of logic. In some books the chapter on thermodynamics contains a laborious geometrical explanation of the partial derivative, although partial derivatives required not even a definition to justify their abundance in the earlier chapters on mechanics. In paraphrasing his old thermodynamics text, after having paraphrased sections from perhaps still older treatises on continuum mechanics, our interdisciplinary compiler-author has forgotten that thermodynamics books are written for a different class of readers, and he reproduces vague, descriptive, and unmathematical wordplay which he himself would not for a moment tolerate in his own chapter on elasticity.

The difference is that thermodynamics never grew up. While aspects of mechanics were developed, applied, generalized, and recast by nearly every distinguished mathematician from ARCHIMEDES to G. D. BIRKHOFF, the unfortunate who reads about thermodynamics even today is made to follow KELVIN's preference for differentials, which KELVIN himself maintained in mechanics as well, but hydrodynamicists and elasticians have long since abandoned, and to suffer over again the insecurity CLAUSIUS seems to have felt whenever he used calculus. For example, our poor reader is expected to believe that something different can come out of a line integral if the line is approximated by infinitesimal adiabatics and isotherms, in defiance of the fundamental theorem of integral calculus.

In this lecture I will show you that classical thermodynamics can be stated precisely and learned, just as classical mechanics is stated precisely and learned. There are problems to be solved in thermodynamics. We can state these problems, and we can solve them in important special cases; more general ones remain to be studied. The range of intended application of classical thermodynamics is to homogeneous systems, namely, bodies that can be described sufficiently by functions of time only. A standard example is furnished by a mixture of fluids in a beaker sufficiently stirred that there are no differences of temperature, concentrations, or other parameters from point to point at any given time. Opening at random a recent and widely used book† on physical chemistry, I found an elegant diagram of just such a beaker, called a "stirred flow reactor",

† Since the example is selected as being typical, not for criticism of any particular author, I do not cite the book quoted.

which is sketched in Figure 1. According to the author, "the reactants enter the vessel at A, and stirring at 3,000 rpm effects mixing within about a second." He tells us that the product mixture is removed at B at a rate exactly balancing the feed, and that after a steady state is attained, the composition of the mixture in the reactor remains unchanged as long as the composition and rate of supply of reactants are unchanged. The numerical values, irrespective of the size of the vessel and of what be the reactants, seem intended to help us keep our feet firmly on the ground of empirical science. I cannot help wishing to see some of my classically thermodynamic friends try the experiment with asphalt as one ingredient and nitroglycerin as the other. Were some mathematician to dream up this apparatus, his colleagues in the natural sciences would rise with con-

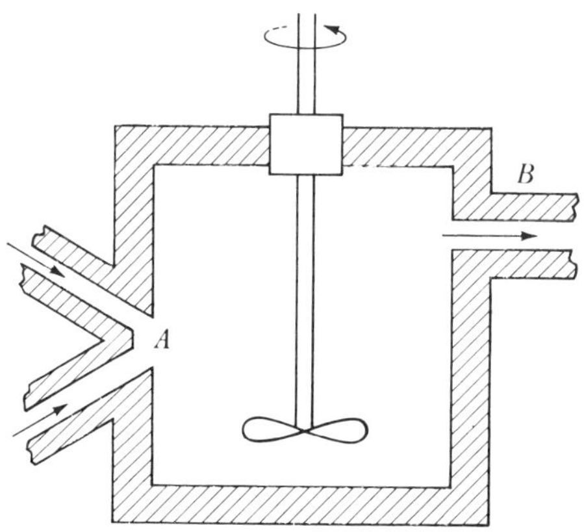

FIGURE 1. Stirred flow reactor.

descending smiles or scathing comments on persons who live in the clouds, but in fact it seems to furnish the standard way to connect classical thermodynamics with the world of experience, although, of course, conditions need not be steady in the theory we shall consider now. This theory describes the same situations, though possibly unsteady, as those for which classical thermodynamics is intended. If the one is applicable, so is the other. All quantities I shall write down will be functions of time, not varying in space, and a dot will denote the time derivative.

I shall draw parallels to classical mechanics. In this instance, "classical mechanics" will mean the Newtonian mechanics of finite systems of mass-points, called "bodies".

Any branch of mathematical physics is constructed in terms of:

1. A list of *primitive quantities*, not defined except by mathematical properties laid down for them.

2. *Definitions* of other quantities in terms of the primitives.
3. *General axioms* stated as mathematical relations satisfied by the primitives and the defined quantities.
4. Proved *theorems* referring to
   a. The theory as a whole, or
   b. Mathematically defined special cases.

The axioms are regarded as *general principles* or *physical laws:* They refer to all systems covered by the theory. Particular systems may be selected and studied at random, but it is more useful to specify them by *constitutive relations* intended to represent important classes of systems. The entire class of constitutive relations is delimited by *constitutive axioms,* which are concrete, mathematical statements in terms of the variables entering the general axioms. The general principles express properties *common* to all systems, while the constitutive relations formalize *diversities* among the systems allowed by the general principles. It should be unnecessary to remark that the choice of primitives, definitions, and axioms is not unique, and that any given theory may be constructed in infinitely many different but equivalent ways. (This fact, however, should not be taken as a license, as often it is, especially in thermodynamics, to present a physical theory in a merely suggestive form, with no mathematical structure at all.)

Table 1 contains a possible list of primitives, axioms, and constitutive equations for classical mechanics and classical thermodynamics. The table merely describes the two theories, not pretending to supply a formal axiomatic development of either.

I will now explain the entries. The "classical mechanics" described here is a fairly special one; easily it could be generalized so as to include variable masses, frictional forces, and multiple interactions, but I have kept it special so that every line will be understood without question by anyone who has studied mechanics in any standard book. The system of thermodynamics outlined, on the other hand, is more general than the usual ones for homogeneous processes, and in form it will surely be unfamiliar. This lecture is devoted to presenting it, though not in the order of the entries. To the few mathematical supplements necessary, I will adjoin simple explanations such as are given in a beginning course in mechanics so as to help the student relate the symbols to everyday experience. At the same time, I will recast the mathematical structure into a more elaborate but more familiar one by introducing appropriate definitions and proving some easy theorems.

Classical thermodynamics refers to only one whole body $\mathcal{B}$, not an infinite system of subbodies, so we need not write the argument $\mathcal{B}$ any more. At each time, the body is assigned a real number called the *temperature.* This number is a measure of how hot the body is. Instruments for

TABLE 1

	Mechanics	*Thermodynamics*
Primitives	Bodies X_p, $p = 1, 2, \ldots, n$ Time t Masses $m(X_p) \equiv m_p > 0$ Place $\mathbf{x}$ Force $\mathbf{f}(X_p, t) \equiv \mathbf{f}_p(t)$ Kinetic process (motion) $\mathbf{x} = \mathbf{x}(X_p, t) \equiv \mathbf{x}_p(t)$	Body $\mathcal{B}$ Time t Temperature $\theta > 0$ Parameters $\boldsymbol{\Upsilon} = (\Upsilon_1, \Upsilon_2, \ldots, \Upsilon_k)$ Net Working $\mathfrak{W}(\mathcal{B}, t)$ Heating $\mathfrak{Q}(\mathcal{B}, t)$ Internal energy $\mathfrak{E}(\mathcal{B}, t)$ Heating bound $\mathfrak{B}(\mathcal{B}, t)$ Thermokinetic process $\theta = \theta(\mathcal{B}, t)$, $\boldsymbol{\Upsilon} = \boldsymbol{\Upsilon}(\mathcal{B}, t)$
Definition	Equilibrium: $\mathbf{x}_p(t) = \text{const.}$	Equilibrium: $\theta(t) = \text{const.}$, $\boldsymbol{\Upsilon}(t) = \text{const.}$
General axioms	$\mathbf{f}_p = m_p \ddot{\mathbf{x}}_p$	$\dot{\mathfrak{E}} = \mathfrak{W} + \mathfrak{Q}$
Constitutive axiom (determinism)	$\mathbf{f}_p = \mathbf{g}_p(\mathbf{x}_p, t) + \sum_{q=1}^{n} \mathbf{f}_{pq}(\mathbf{x}_p, \mathbf{x}_q, t)$ where $\mathbf{f}_{pq} = f_{pq} \dfrac{\mathbf{x}_p - \mathbf{x}_q}{\lvert \mathbf{x}_p - \mathbf{x}_q \rvert}$, $f_{pq} = f_{qp}$ in every process	$\mathfrak{W} = \mathbb{W}(\theta^t, \boldsymbol{\Upsilon}^t)$, $\mathfrak{E} = \mathbb{E}(\theta^t, \boldsymbol{\Upsilon}^t)$, $\mathfrak{B} = \mathbb{B}(\theta^t, \boldsymbol{\Upsilon}^t)$, where $\mathfrak{W} = \mathfrak{Q} = 0$ in equilibrium and $\mathfrak{Q} \leqq \mathfrak{B}$ in every process

measuring it are called "thermometers" or "thermocouples". Experiments show that whatever such instrument be used, there is a temperature below which no body can be cooled. This least possible temperature depends on the thermometer used but not on the bodies whose temperature it measures. If we assign to this universal greatest lower bound the value zero, the temperature scale is said to be *absolute*. We need not elevate this remark to the status of an axiom. Rather, as in mechanics the student is accustomed from the start to accepting the mass of a body as a positive quantity, in every system of units and by every method of measurement, we may accept the existence of an absolute temperature for constructing the theory.†

† It may be worthwhile to remark that if we accept a molecular explanation for the temperature as being, in some sense, an average or mean kinetic energy of the particles, then the assumption that all masses are positive implies that all temperatures are non-negative. Thus in motivating the respective mathematical theories, we ought to justify *both* assumptions or *neither*. I choose the latter alternative.

Just as in mechanics a body is located at some place at each time, in thermodynamics a body is described by a list of real parameters Υ_1, Υ_2, . . . , Υ_k, denoted collectively by $\boldsymbol{\Upsilon}$. In the commonest special case, $\boldsymbol{\Upsilon}$ consists in, first, the volume V of the region $\mathcal{V}$ occupied by the body $\mathcal{B}$ and the masses of $k - 2$ of the $k - 1$ chemically or physically distinguishable phases which make up the body. In the favorite case of thermodynamics books, a homogeneous fluid, $\boldsymbol{\Upsilon}$ reduces to a single parameter, the volume V. The problem of selecting and identifying $\boldsymbol{\Upsilon}$ in a way appropriate to some physical body lies outside the theory itself. It is analogous to the problem in applied mechanics of looking at a natural body and deciding how many degrees of freedom should be assigned to a mathematical model for it in the theory. Important as this problem is for applications, it does not form part of the mathematical theory. $\boldsymbol{\Upsilon}$, therefore, remains for us no more than a fixed list of real numbers.

Thermodynamics differs from other branches of mathematical physics only in its emphasis on the effects of heating and change of temperature. Its aims are three:

1. *Reduction of constitutive equations:* To formulate, within a specified class, the most general models of bodies that are consistent with the axioms governing the transfer and dissipation of energy, and to discover the properties of those models.
2. *Construction of the equilibrium theory:* From the general principles governing the evolution of thermodynamic processes in time, to discover and demonstrate the principles of thermostatics, that is, the properties that distinguish an actual equilibrium configuration within the class of all putative ones.
3. *Application:* When the actions of the outside world on a specified body are prescribed, to determine the process that the body must undergo so as to support them.

In these lectures we shall not pursue the third aim at all. For illustration of work toward it, extremely special, the reader may consult any of the standard treatises on the conduction of heat or the diffusion of mass. In this course, my purpose is to explain by example how the first and second aims may be reached. In this, the first lecture, we consider only the most elementary case.

A *thermokinetic process* is the pair of functions† assigning θ and $\boldsymbol{\Upsilon}$ to

† In these lectures I often use the same symbol to denote a function and its value. When the distinction is to be emphasized, $\theta(\cdot)$ will be written for the function whose value is θ. Sometimes, however, I shall use a different letter for the function, for example, $y = f(x)$. There is then no need to write $f(\cdot)$ for the function f.

$\mathcal{B}$ at each time t:

$$\theta = \theta(t) > 0, \qquad \Upsilon = \Upsilon(t). \tag{1.5}$$

For the present, we shall assume that $\mathcal{B}$ is unconstrained in the sense that it may experience any continuously differentiable thermokinetic process.

Thermodynamics presupposes some mechanical theory for each body to which it is applied. In mechanics the concepts of work, power, and kinetic energy are defined. By "mechanics" at this moment I refer to something more general than the case set up for analogy in the left-hand column of Table 1. Typically, some rudiments of some branch of continuum mechanics such as gas dynamics are applied by thermodynamicists, if often tacitly, as part of the specification of their systems. The *net working* $\mathfrak{W}$ at time t on the body $\mathcal{B}$ is then given by

$$\mathfrak{W} \equiv \mathfrak{P} - \dot{\mathfrak{K}}, \tag{1.6}$$

where $\mathfrak{P}$ is the mechanical power and $\mathfrak{K}$ is the kinetic energy. Thus $\mathfrak{W}$ is the power not used up in producing motion. Often $\dot{\mathfrak{K}}$ is neglected. For our purposes it is sufficient to say that $\mathfrak{W}$, like Υ_a, is left an arbitrary real-valued function of time, to be identified in various ways by means of experience in mechanics. The physical dimensions of $\mathfrak{W}$ are those of power, namely, force times distance per unit time.

If we recognize that doing work may change the energy of a body, and that heating may effect work, at this level of generality the best we can say is that to describe this *equivalence of heat and work,* we assume the existence of a second kind of working, $\mathfrak{Q}$, called the *heating,* which is not identified with anything from mechanics. $\mathfrak{Q}$ is a function of time having the same dimensions as $\mathfrak{W}$ and hence may be added to it so as to yield the rate of increase of *internal energy* $\mathfrak{E}$ according to the ***first axiom:***

$$\dot{\mathfrak{E}} = \mathfrak{W} + \mathfrak{Q}, \tag{1.7}$$

often called "the first law of thermodynamics".

(Just as the "Newtonian" equation $\mathbf{f}_p = m_p\ddot{\mathbf{x}}_p$ looks like a mathematical definition of force and has been mistakenly regarded as one by many authors, so also (1.7) looks like, but is not, a definition of internal energy $\mathfrak{E}$. The internal energy of $\mathcal{B}$, like the force acting on X_p, is a quantity about which we have *a priori* knowledge in any particular case. Both represent actions of the surroundings upon the bodies. $\mathfrak{Q}$ is introduced now for later use, and it will enter another axiom below. Thus the "first law", like the "Newtonian equations", is a restriction upon the quantities entering it but not a definition of any one of them.)

The irreversibility of natural processes is represented by the existence of an *a priori* least upper bound $\mathfrak{B}$ for the heating $\mathfrak{Q}$. The term "irreversibility" is justified because $\mathfrak{Q}$, the rate of increase of energy not accompanied by mechanical working, is bounded above but not necessarily below. There is no limit to the magnitude of negative values of $\mathfrak{Q}$, which represent conversion of energy into heating without performance of work, but there certainly is to positive ones. Work and energy may always be converted into heat, but there is a limit to the rate at which heat may be converted into energy without doing mechanical work. This is the content of the ***second axiom:***

$$\mathfrak{Q} \leqq \mathfrak{B}, \tag{1.8}$$

or, equivalently, by (1.7),

$$\dot{\mathfrak{E}} - \mathfrak{W} \leqq \mathfrak{B}. \tag{1.9}$$

A *calorodynamic process* for $\mathcal{B}$ is a set of four functions of time

$$\mathfrak{W}(t), \qquad \mathfrak{E}(t), \qquad \mathfrak{Q}(t), \qquad \mathfrak{B}(t), \tag{1.10}$$

that satisfy the axioms of thermodynamics, namely, (1.7) and (1.8).

While the greatest lower bound or zero of temperature θ is universal, for all bodies and all times, the upper bound $\mathfrak{B}$ for the heating $\mathfrak{Q}$ depends not only on the time but also on the nature of the body. In one body it may be possible to store internal energy faster by heating without working than it is for some other body in the same circumstances. Thus the heating bound $\mathfrak{B}$, like the forces $\mathbf{f}_p$ in mechanics, is a constitutive quantity. So are $\mathfrak{W}$ and $\mathfrak{Q}$ and $\mathfrak{E}$. That is, the diversity of bodies is represented in thermodynamics by *constitutive relations* for $\mathfrak{W}$, $\mathfrak{Q}$, $\mathfrak{E}$, and $\mathfrak{B}$. The constitutive relations embody the ***constitutive axiom*** or *principle of thermodynamically compatible determinism.* Namely, for a given body $\mathcal{B}$ at time t, *the history of the thermokinetic process to which it has been subject up to and including the time t determines a calorodynamic process for it.*

To make this idea clearer, we define f^t, the *history* of the function f up to time t, as the function cut off at the present time:

$$f^t(s) \equiv f(t - s), \qquad 0 \leqq s < \infty. \tag{1.11}$$

Then the constitutive axiom asserts that for a given body there exist functionals $\mathbb{W}$, $\mathbb{Q}$, $\mathbb{E}$, $\mathbb{B}$ such that

$$\begin{aligned} \mathfrak{W}(t) &= \mathbb{W}(\theta^t, \mathbf{\Upsilon}^t), \\ \mathfrak{Q}(t) &= \mathbb{Q}(\theta^t, \mathbf{\Upsilon}^t), \\ \mathfrak{E}(t) &= \mathbb{E}(\theta^t, \mathbf{\Upsilon}^t), \\ \mathfrak{B}(t) &= \mathbb{B}(\theta^t, \mathbf{\Upsilon}^t), \end{aligned} \tag{1.12}$$

and moreover, that these four functionals are such as to satisfy (1.7) and (1.8) *for every thermokinetic process.* A set of four functionals $\mathbb{W}$, $\mathbb{Q}$, $\mathbb{E}$, $\mathbb{B}$ of this kind defines a *thermodynamic body* $\mathcal{B}$.

The requirement of compatibility with (1.7) may be satisfied at once and trivially by regarding the functional $\mathbb{Q}$ as determined by $\mathbb{E}$ and $\mathbb{W}$.

In effect, we simply omit $\mathfrak{Q}$ henceforth from the list of quantities given by constitutive relations.

From now on, the term *thermodynamic process* or simply *process* will refer to a thermokinetic process or a calorodynamic process or both, on the understanding that they are related through the constitutive axiom.

We shall see now that these ideas are less unfamiliar than they may seem. First, since $\theta > 0$, we are justified in defining the *calory* $\mathfrak{H}$ produced in the process from t_0 to t:

$$\mathfrak{H}(t, t_0) \equiv \int_{t_0}^{t} \frac{\mathfrak{B}}{\theta}\, dt. \tag{1.13}$$

Since $\mathfrak{B} = \theta\dot{\mathfrak{H}}$, the second axiom (1.8) now becomes

$$\theta\dot{\mathfrak{H}} \geqq \mathfrak{Q}. \tag{1.14}$$

If we call $\mathfrak{H}$ the "entropy", we see that the second axiom is an expression in terms of time rates of what is often called "the second law of thermodynamics", in the Clausius-Planck form. Since $\mathfrak{H}$ is defined only to within an additive constant, other quantities defined from $\mathfrak{H}$ will reflect this indeterminacy; likewise, by (1.7), a constant may be added to $\mathfrak{E}$ without affecting any property yet laid down for it. While any result intended for comparison with experience must remain unchanged under the addition of these two arbitrary constants, in these lectures I shall not stop to verify that such is the case.

(I hesitate to use the terms "first law" and "second law", because there are almost as many "first and second laws" as there are thermodynamicists, and I have been told by these people for so many years that I disobey their laws that now I prefer to exult in my criminal status and give non-condemning names to the concrete mathematical axioms I wish to use in my outlaw studies of heat and temperature. The term "entropy" seems superfluous, also, since it suggests nothing to ordinary persons and only intense headaches to those who have studied thermodynamics but have not given in and joined the professionals. Since the reasons for which, about a century ago, the words "heat" and "caloric" were banished from physics except in attributive or pejorative senses, now have only historical interest, we should be allowed to bring their suggestive power back into the formal theory. In so doing we circumvent the persons who claim that "entropy cannot be defined except for systems infinitesimally near to equilibrium." The reasons they give are of two kinds: (1) some definition or other from statistical mechanics is well motivated only in statistical equilibrium, and (2) professional authority. Since here we make no appeal to statistical mechanics or to authority, use of a name other than "entropy" neatly sets aside both these objections.)

Of course, the calory $\mathfrak{H}$ should not be confused with the *heat gained* by

$\mathfrak{B}$ in the same process, which is $\int_{t_0}^{t} \mathfrak{Q}\,dt$ and hence is an energy, while $\mathfrak{H}$ is an energy per unit temperature. The two are related by the evident inequality

$$\int_{t_0}^{t} \mathfrak{Q}\,dt \leqq \theta\mathfrak{H}\Big|_{t_0}^{t} - \int_{t_0}^{t} \dot{\theta}\mathfrak{H}\,dt. \tag{1.15}$$

A process is called *reversible* if for all t the sign "$=$" holds in (1.8) (and hence in (1.9), (1.14), and (1.15)). All other processes are called *irreversible*. These terms are most easily motivated by special cases. A process is

$$\left.\begin{array}{l} \textit{adiabatic} \\ \textit{isocaloric} \\ \textit{isothermal} \end{array}\right\} \text{if for all } t \left\{\begin{array}{l} \mathfrak{Q} = 0 \\ \mathfrak{B} = 0 \\ \theta = \text{const.} \end{array}\right. \tag{1.16}$$

In the foregoing definitions "all t" is understood as meaning "all but a set of measure 0." Thus we have the following mathematically trivial statements:

$\mathfrak{Q} = 0 \Rightarrow \dot{\mathfrak{E}} = \mathfrak{W}\ \&\ \dot{\mathfrak{H}} \geqq 0.$ In an adiabatic process, the energy gained is the work done. In a reversible adiabatic process, the calory is constant. In an irreversible adiabatic process, the calory increases.

$\dot{\mathfrak{H}} = 0 \Rightarrow \dot{\mathfrak{E}} \leqq \mathfrak{W}\ \&\ \mathfrak{Q} \leqq 0.$ A reversible isocaloric process is adiabatic. In an irreversible isocaloric process, the energy gained is less than the work done, and some heat is lost.

$\dot{\theta} = 0 \Rightarrow \mathfrak{H}(t) - \mathfrak{H}(t_0) \geqq \dfrac{1}{\theta}\displaystyle\int_{t_0}^{t} \mathfrak{Q}\,dt.$ In a reversible isothermal process, the increase of calory is the heat gained per unit absolute temperature. In an irreversible isothermal process, the increase of calory is greater than in a reversible process at the same temperature.

More sophisticated inequalities may be drawn from (1.15) by the aid of the mean-value theorems of calculus.

For the mathematical development it helps to introduce the *free energy* $\mathfrak{F}$ by the classical definition†:

$$\mathfrak{F} \equiv \mathfrak{E} - \theta\mathfrak{H}. \tag{1.17}$$

† As remarked above, $\mathfrak{E}$ and $\mathfrak{H}$ are indeterminate to within additive constants. Thus the physical interpretation of any result should be insensitive to addition of a linear function of θ to $\mathfrak{F}$.

Then

$$\begin{aligned} \dot{\mathfrak{F}} &= \dot{\mathfrak{E}} - \mathfrak{H}\dot{\theta} - \theta\dot{\mathfrak{H}}, \\ &= \mathfrak{W} - \mathfrak{H}\dot{\theta} + \mathfrak{Q} - \theta\dot{\mathfrak{H}} \end{aligned} \tag{1.18}$$

by (1.7). Hence the Clausius-Planck inequality (1.14) assumes the form called the ***reduced dissipation inequality:***

$$\dot{\mathfrak{F}} - \mathfrak{W} + \mathfrak{H}\dot{\theta} \leqq 0. \tag{1.19}$$

The term "reduced" refers to the fact that $\mathfrak{Q}$ has been eliminated by use of (1.7).

By (1.13), $\mathfrak{B} = \theta\dot{\mathfrak{H}}$, so $(1.12)_4$ asserts that the function $\dot{\mathfrak{H}}$ is the value of a functional of the thermokinetic process. Consequently, so also is the function $\mathfrak{H}$. By (1.17) and $(1.12)_3$, then, so also is $\mathfrak{F}$. All four functions occurring in (1.19) are thus values of functionals of the thermokinetic process, and we may obtain from (1.17) the following ***effective principle of thermodynamic determinism:***

$$\begin{aligned} \mathfrak{W} &= \mathbb{W}(\theta^t, \boldsymbol{\Upsilon}^t), \\ \mathfrak{F} &= \mathbb{P}(\theta^t, \boldsymbol{\Upsilon}^t), \\ \mathfrak{H} &= \mathbb{H}(\theta^t, \boldsymbol{\Upsilon}^t), \end{aligned} \tag{1.20}$$

where the functionals $\mathbb{W}$, $\mathbb{P}$, *and* $\mathbb{H}$ *are such that the reduced dissipation inequality* (1.19) *is satisfied identically in the history* θ^t, $\boldsymbol{\Upsilon}^t$, *provided only* $\dot{\theta}$ *and* $\dot{\mathfrak{F}}$ *exist.* Conversely, if the functionals in (1.20) are such that (1.19) is satisfied whenever $\dot{\theta}$ and $\dot{\mathfrak{F}}$ exist, we may retrace our steps and recover the constitutive axiom (1.12), provided the value $\mathfrak{H}(t)$ of the functional $\mathbb{H}$ be differentiable. Three functionals $\mathbb{W}$, $\mathbb{P}$, and $\mathbb{H}$ that satisfy this axiom will be called *constitutive functionals of a thermodynamic body.* In view of the various connections with older ideas we have developed in the foregoing paragraphs, we can interpret this new formulation, equivalent to our original one (except for cases in which $\dot{\mathfrak{F}}$ and $\dot{\mathfrak{H}}$ fail to exist), as follows: *A thermodynamic body is defined by a rule which determines the present working, present free energy, and present calory from the thermokinetic history* θ^t, $\dot{\boldsymbol{\Upsilon}}^t$ *in such a way that the reduced dissipation inequality holds for every process. Conversely, any such rule defines a thermodynamic body.* In rougher terms, a body must be such that we cannot imagine any way to make it disobey the "first and second laws of thermodynamics". That is, the "first and second laws" are to be interpreted, not as restrictions on the processes a body must undergo, but as restrictions on the response of the body itself†.

† Ever since the new thermodynamics was opened by COLEMAN & NOLL in 1963, the secondary literature has registered confusion about it. The typical complaint, reformulated in terms of the variables used in this lecture, has been, "How can the body heating $\mathfrak{Q}$ and the thermokinetic history θ^t, $\boldsymbol{\Upsilon}^t$ be arbitrary, when in the

We can now state the ***first problem of thermodynamics:*** *in an assigned class of thermokinetic processes and an assigned class of functionals* $\mathbb{W}$, $\mathbb{P}$, $\mathbb{H}$, *to determine all constitutive functionals.* It is understood that for each t in $(-\infty, \infty)$ the argument θ^t is a positive-valued function on $[0, \infty)$ and the arguments $\Upsilon_1{}^t, \ldots, \Upsilon_k{}^t$ are real-valued functions on $[0, \infty)$. To get concrete results, it is generally necessary to restrict the argument functions to some particular function-space, as the statement suggests.

Thus the axiom (1.8) is not so trivial as it looks. Of course we can satisfy it by the choice $\mathfrak{B} = \mathfrak{Q}$, but the problem is to find *all* possible choices *in a given class of functionals over a given class of arguments.* Those classes may be such as to make $\mathfrak{B} = \mathfrak{Q}$ the only choice, one of many

---

applications we assign $\mathfrak{Q}$ and determine θ^t and $\boldsymbol{\Upsilon}^t$ from initial and boundary data?" The question itself reflects a confusion in the mind of the questioner, a confusion natural enough even today in view of the ingrained custom of mixing up constitutive relations with general physical laws. The one theory where this confusion has been avoided or corrected is the Newtonian doctrine of mass-points, where no one would dream of saying, "What are these arbitrary forces? In nature we encounter only gravitational, elastic, and electromagnetic forces, so anything further is unphysical." In fact, the mechanics of mass-points was established generally and correctly long before the nature of electromagnetic forces was determined, so such a claim would run against history as well as be stupid for reasons of method and mathematics.

Rather than refute current objections, which in any case are inchoate if not morbid, I shall merely state clearly the position I take here:

1. In considering everything that might *ever* happen to *any possible body*, there is no reason to exclude any particular thermokinetic history θ^t, $\boldsymbol{\Upsilon}^t$ or any particular body heating $\mathfrak{Q}$. Therefore, when we seek overriding restrictions on *all possible constitutive relations of bodies,* these functions must be left arbitrary within some function space, the generality of which is limited only by the present resources of mathematicians and mathematical analysis.
2. Once the constitutive relations are determined, we turn to the modelling of particular experiments on the bodies those relations are intended to describe. For these experiments, of course, the body heating is generally assigned, not arbitrary. Prescription of $\mathfrak{Q}$ and subsidiary quantities such as the initial value of $\mathfrak{E}$ amounts to a restriction on the thermokinetic process θ^t, $\boldsymbol{\Upsilon}^t$, and determination of all such processes (perhaps reducing to a single one) is a typical problem of the older kind. Such problems *cannot even be stated* until the constitutive relations have been fixed.

It should be unnecessary to remark, although unfortunately it seems not to be, that in the first analysis there is nothing inconsistent with or repugnant to the second one, or *vice versa.* Those who require a classical precedent should turn to the Navier-Stokes theory of fluids, in which, as will be made clear in the next lecture, the first analysis in rudimentary form has *always* been applied so as to prove that the shear viscosity and bulk viscosity must be non-negative, which facilitates rather than impedes the solution of important special problems in analyses of the second kind.

allowed choices, or altogether forbidden. We shall illustrate these three possibilities by presenting the two important special cases in which the problem has been solved, in effect, in the older literature.

Example 1. *The classical "equations of state"*. The assigned class of functionals is an extremely simple one. Namely, the free energy and the calory are supposed to be determined by the present state, irrespective of past states, and the working, which is itself a rate, is taken to be a homogeneous linear function of the rates of change $\dot{\Upsilon}_a$ of the parameters Υ_a:

$$\begin{aligned} \mathfrak{W} &= -\sum_{a=1}^{k} \varpi_a(\theta, \boldsymbol{\Upsilon})\dot{\Upsilon}_a, \\ \mathfrak{F} &= \Psi(\theta, \boldsymbol{\Upsilon}), \\ \mathfrak{H} &= \mathrm{H}(\theta, \boldsymbol{\Upsilon}) \end{aligned} \tag{1.21}$$

This special constitutive class represents materials totally devoid of memory. The coefficients $\varpi_a(\theta, \boldsymbol{\Upsilon})$ are called *thermodynamic pressures*, and it is to be emphasized that they, like Ψ and H, are functions of θ and $\boldsymbol{\Upsilon}$ alone. Ψ is assumed to be differentiable, and the class of thermokinetic processes is assumed to include all continuously differentiable ones. Because of the effective principle of determinism, the $k + 2$ functions of state ϖ_a, Ψ, and H in order to define a thermodynamic body must satisfy the reduced dissipation inequality (1.19) for every process θ^t, $\boldsymbol{\Upsilon}^t$. Hence

$$(\partial_\theta \Psi + \mathrm{H})\dot{\theta} + \sum_{a=1}^{k} (\partial_{\Upsilon_a}\Psi + \varpi_a)\dot{\Upsilon}_a \leqq 0 \tag{1.22}$$

for arbitrary positive $\theta(t)$ and arbitrary real $\dot{\theta}(t)$, $\Upsilon_a(t)$, and $\dot{\Upsilon}_a(t)$. Since the coefficients of $\dot{\theta}$ and $\dot{\Upsilon}_a$ in (1.22) are functions of θ and $\boldsymbol{\Upsilon}$ only, at each fixed θ and $\boldsymbol{\Upsilon}$ that inequality requires a homogeneous linear function of $\dot{\theta}$ and ϖ to be of one sign. This is possible if and only if that linear function vanishes identically. Hence

$$\begin{aligned} \mathrm{H} &= -\partial_\theta \Psi, \\ \varpi_a &= -\partial_{\Upsilon_a}\Psi. \end{aligned} \tag{1.23}$$

We have proved these conditions to be necessary. Substituting them back into (1.22) shows not only that they are sufficient as well but also that "=" holds always and "<" never. Thus we have the following theorem, phrased in semiclassical terms: *The caloric equations of state* (1.21) *define a thermodynamic body if and only if the* $k + 1$ *functions* ϖ_a *and* H *are determined from the free-energy function* Ψ *as a "thermodynamic potential". In the body so defined, all processes are reversible, and a process is adiabatic if and only if it is isocaloric.*

This theorem, along with changes rung on it by transformation to other potentials and corollaries obtained from it by differentiation and integration, comes close to exhausting the mathematical contents of the standard books on thermodynamics. One corollary important conceptually should be mentioned here. The starting assumption (1.21) suggests that we might have to make a measurement in a non-static process ($\dot{\mathbf{\Upsilon}} \neq \mathbf{0}$) in order to determine the thermodynamic pressure $\varpi_a(\theta, \mathbf{\Upsilon})$ by experiment. The result (1.23) shows this not to be so. *The function* $\Psi(\theta, \mathbf{\Upsilon})$, *determinable by equilibria alone, determines in turn all the functions occurring in the equations of state.* Our presentation makes it clear that the most prominent and famous features of the classical theory, namely, the existence of the classical potentials and the consequent reduction of the content of thermodynamics to statements about equilibrium, *i.e.*, to a part of *thermostatics*, is not at all essential to the general principles of the subject but instead is a consequence of assuming a particularly simple constitutive class at the outset.

Remark on Example 1. In the analysis of Example 1 we have assumed not only that the constitutive function Ψ was differentiable but also that the class of thermokinetic processes included *all* differentiable functions $\theta(\cdot)$, $\mathbf{\Upsilon}(\cdot)$. The same results can be obtained under the weaker assumptions on the processes and slightly stronger assumptions on the functions. Namely, let Ψ be assumed continuously differentiable, and H and ϖ_a continuous. Let the processes, defined over $(-\infty, +\infty)$, be restricted to some class $\mathcal{C}$, of which we require only that it be *closed under local linear continuations.* The local linear continuation f_u of a function f at $t = t_0$ is defined as follows:

$$f_u(t) \equiv \begin{cases} f(t), & -\infty < t < t_0, \\ f(t_0) + (t - t_0)u, & t_0 \leqq t < t_0 + k, \end{cases} \tag{1.24}$$

where u is a real number and k is a positive number. The diagram shows the local linear continuation for a case when the function f has a non-removable discontinuity at t_0 (see Figure 2). The assumption we make on the class $\mathcal{C}$ of thermokinetic processes $\theta(\cdot)$, $\mathbf{\Upsilon}(\cdot)$ is that if $\theta(\cdot)$, $\mathbf{\Upsilon}(\cdot)$ is in $\mathcal{C}$, so also are *all* local linear continuations, for sufficiently small k. That is, given any time t_0 and any real numbers u and v_a, we can find a positive k such that θ_u, $\mathbf{\Upsilon}_{\mathbf{v}}$ is also† in $\mathcal{C}$. This assumption provides us with a smooth process going forward for some time, perhaps only a short one, from any state for which the constitutive functions are defined. It enables us to dispense with assumptions of smoothness on the processes at $t = t_0$. The class $\mathcal{C}$ need not contain any process that is even piecewise continuous in every

† In order to enter a thermokinetic process, by definition $\theta_u(t)$ must be positive. Since $\theta(t)$ does enter a thermokinetic process, $\theta(t_0) > 0$. Hence for sufficiently small k the linear continuation θ_u will always be positive, no matter what u is.

sufficiently small interval† about t_0. (The reduced dissipation inequality (1.19), of course, is required to hold only at points where it has meaning, namely, where the thermokinetic process is differentiable.) In return for this freedom to restrict strongly the class of processes for which the theory is formulated, we agree to consider only such functions ϖ_a, $\partial_\theta \Psi$, $\partial_{\Upsilon_a}\Psi$, and H as are continuous at every state corresponding to any process in $\mathcal{C}$.

We fix the time t_0 and let $\theta(\cdot)$, $\boldsymbol{\Upsilon}(\cdot)$ be a process in $(-\infty, t_0)$. For the

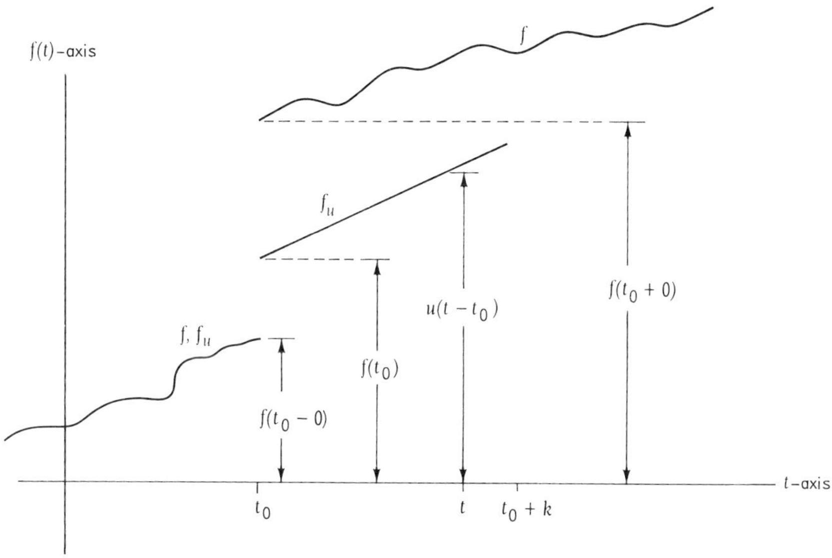

FIGURE 2. Local linear continuation of f at t_0.

local linear continuation θ_u, $\boldsymbol{\Upsilon}_{\mathbf{v}}$ of that process, we have for some positive k

$$\dot{\Psi} = (\partial_\theta \Psi)u + \sum_{a=1}^{k} (\partial_{\Upsilon_a}\Psi)v_a, \qquad t_0 < t < t_0 + k. \tag{1.25}$$

Since the reduced dissipation inequality (1.19) must hold for every differentiable process in $\mathcal{C}$, it must hold in particular for the local linear continuation θ_u, $\boldsymbol{\Upsilon}_{\mathbf{v}}$ when $t_0 < t < t_0 + k$. Hence for any fixed u and $\mathbf{v}$,

$$(\partial_\theta \Psi + \mathrm{H})_t u + \sum_{a=1}^{k} (\partial_{\Upsilon_a}\Psi + \varpi_a)_t v_a \leqq 0. \tag{1.26}$$

As indicated by the subscript t, the arguments of the functions forming

† For given f, the positive constant k is a function of t_0 and u. For each given t_0 and u, we assume that a positive k exists, but as $t_0 \to \infty$ or $u \to \pm\infty$, the greatest lower bound of k need not be positive.

the coefficients of u and v_a are

$$\theta(t_0) + (t - t_0)u, \qquad \Upsilon_a(t_0) + (t - t_0)v_a.$$

But, by assumption, those coefficient functions are continuous. Hence as $t \to t_0 + 0$, (1.26) yields

$$(\partial_\theta \Psi + \mathrm{H})_{t_0} u + \sum_{a=1}^{k} (\partial_{\Upsilon_a} \Psi + \varpi_a)_{t_0} v_a \leqq 0, \tag{1.27}$$

where, as indicated by the subscript t_0, the arguments are now $\theta(t_0)$, $\boldsymbol{\Upsilon}(t_0)$. Since u and v_a are arbitrary real numbers, (1.23) follows for the state $\theta(t_0)$, $\boldsymbol{\Upsilon}(t_0)$, but since that state may be any one assumed by a process in $\mathfrak{C}$, (1.23) holds in general.

Example 2. *"Linear friction"*. To include effects commonly described as those of linear friction, it is customary to retain the caloric equations of state for $\mathfrak{F}$ and $\mathfrak{H}$ but take $\mathfrak{W}$ to be given by a quadratic function of $\dot{\boldsymbol{\Upsilon}}$:

$$\begin{aligned} \mathfrak{W} &= -\sum_{a=1}^{k} \varpi_a{}^0(\theta, \boldsymbol{\Upsilon})\dot{\Upsilon}_a - \sum_{a,b=1}^{k} \varpi_{ab}(\theta, \boldsymbol{\Upsilon})\dot{\Upsilon}_a\dot{\Upsilon}_b, \\ \mathfrak{F} &= \Psi(\theta, \boldsymbol{\Upsilon}), \\ \mathfrak{H} &= \mathrm{H}(\theta, \boldsymbol{\Upsilon}). \end{aligned} \tag{1.28}$$

Instead of (1.22) we now obtain the more general inequality

$$(\partial_\theta \Psi + \mathrm{H})\dot{\theta} + \sum_{a=1}^{k} (\partial_{\Upsilon_a}\Psi + \varpi_a{}^0)\dot{\Upsilon}_a + \sum_{a,b=1}^{k} \varpi_{ab}\dot{\Upsilon}_a\dot{\Upsilon}_b \leqq 0. \tag{1.29}$$

Since the coefficients of $\dot{\theta}$ and $\dot{\Upsilon}_a$ and $\dot{\Upsilon}_b\dot{\Upsilon}_c$ are functions of θ and $\boldsymbol{\Upsilon}$ only, this inequality requires that a paraboloid $y = \Sigma a_i x_i + \Sigma a_{ij} x_i x_j$ lie entirely on one side of the plane $y = 0$. Such a paraboloid must have its axis normal to the plane $y = 0$ and be elliptic. Hence (1.29) is satisfied if and only if for every θ and $\boldsymbol{\Upsilon}$

$$\begin{aligned} \mathrm{H} &= -\partial_\theta \Psi, \\ \varpi_a{}^0 &= -\partial_{\Upsilon_a}\Psi, \\ \sum_{a,b=1}^{k} \varpi_{ab} Y_a Y_b &\leqq 0 \qquad \text{for all } \mathbf{Y}. \end{aligned} \tag{1.30}$$

Accordingly, in a thermodynamic body with linear friction the free-energy function Ψ is a potential for H and $\varpi_a{}^0$ only, but imposes no conditions on ϖ_{ab}. Thus the free-energy function no longer determines all the thermodynamic properties of the body. The coefficients ϖ_{ab} are not entirely arbitrary, however, since their quadratic form must be negative-semidefinite. Finally, for the material with linear friction, irreversible processes are possible and indeed typical, and adiabatic processes can no longer be confounded with isocaloric ones.

This second example allows us to illustrate the nature of the heating bound $\mathfrak{B}$. Indeed, by (1.13), and (1.18), and $(1.28)_1$ we find that

$$\begin{aligned} \mathfrak{B} - \mathfrak{Q} &= \mathfrak{W} - \dot{\mathfrak{F}} - \dot{\theta}\mathfrak{H}, \\ &= - \sum_{a,b=1}^{k} \varpi_{ab}(\theta, \boldsymbol{\Upsilon})\dot{\Upsilon}_a\dot{\Upsilon}_b. \end{aligned} \tag{1.31}$$

If $\varpi_{ab} \equiv 0$ for all a and b, then $\mathfrak{B} = \mathfrak{Q}$ always. If $\|\varpi_{ab}\|$ is negative-definite, $\mathfrak{B} = \mathfrak{Q}$ if and only if the process is static. That is, only static processes are reversible in such a body. If $\|\varpi_{ab}\|$ is negative-semidefinite but neither definite nor the zero matrix, then $\mathfrak{B} = \mathfrak{Q}$ in some non-static processes but not in others; that is, for such a body, some non-static processes are reversible and some are not. Thus all three possible relations between the heating $\mathfrak{Q}$ and its constitutive least upper bound $\mathfrak{B}$ are illustrated by the bodies included in Example 2.

The second example includes the first as a special case. These are only the simplest classes of materials for which the problem of thermodynamics has been solved. In the third lecture, a far greater class will be found.

Further Remarks on the Examples. The classical assumption (1.21) does not satisfy the *principle of equipresence,* which forbids the theorist from choosing independent variables by caprice. According to that principle, we should replace (1.21) by

$$\begin{aligned} \mathfrak{W} &= W(\theta, \boldsymbol{\Upsilon}) - \sum_{a=1}^{k} \varpi_a(\theta, \boldsymbol{\Upsilon})\dot{\Upsilon}_a, \\ \mathfrak{F} &= \Psi(\theta, \boldsymbol{\Upsilon}) - \sum_{a=1}^{k} \pi_a(\theta, \boldsymbol{\Upsilon})\dot{\Upsilon}_a, \\ \mathfrak{H} &= \mathrm{H}(\theta, \boldsymbol{\Upsilon}) - \sum_{a=1}^{k} \eta_a(\theta, \boldsymbol{\Upsilon})\dot{\Upsilon}_a, \end{aligned} \tag{1.32}$$

so that the independent variables and analytic restrictions imposed at the outset are the same for all three constitutive functionals. According to the requirement that $\mathfrak{W} = 0$ in equilibrium (listed in Table 1 but not otherwise mentioned up to now), $W(\theta, \boldsymbol{\Upsilon}) \equiv 0$. If we substitute (1.32) into the reduced dissipation inequality (1.19), we obtain instead of (1.22)

$$\begin{aligned} \Big[\partial_\theta\Psi + \mathrm{H} - \sum_{a=1}^{k} (\eta_a + \partial_\theta\pi_a)\dot{\Upsilon}_a\Big]\,\dot{\theta} + \sum_{a=1}^{k} (\partial_{\Upsilon_a}\Psi + \varpi_a)\dot{\Upsilon}_a \\ - \sum_{a,b=1}^{k} (\partial_{\Upsilon_a}\pi_b)\dot{\Upsilon}_a\dot{\Upsilon}_b - \sum_{a=1}^{k} \pi_a\ddot{\Upsilon}_a \leqq 0. \end{aligned} \tag{1.33}$$

As we have seen in treating Example 2, an inequality of this kind can be satisfied identically if and only if the coefficients of the linear terms vanish identically, while the remaining quadratic form is negative-semidefinite. The first condition requires, among other things, that $\pi_a \equiv 0$; the second thus reduces to

$$-\left(\sum_{a=1}^{k} \eta_a \dot{\Upsilon}_a\right) \dot{\theta} \leqq 0, \tag{1.34}$$

which requires that $\eta_a \equiv 0$. Thus the classical results (1.23) follow also from the apparently more general starting assumption (1.32).

Likewise, in the example concerning linear friction, we can add terms both linear and quadratic in $\dot{\mathbf{\Upsilon}}$ to the forms $(1.28)_{2,3}$ assumed for $\mathfrak{F}$ and $\mathfrak{H}$ at the outset. By an easy analysis again we recover (1.30). Thus the classical theory of "linear friction" also squares with the principle of equipresence.

To get something different from the two classical special cases, we need to depart further from their starting assumptions; for example, we could put terms in $\dot{\theta}$ or $\ddot{\mathbf{\Upsilon}}$ into $\mathfrak{W}$, or introduce long-range memory, *e.g.*,

$$\mathfrak{F} = \Psi^0(\theta(t), \mathbf{\Upsilon}(t)) + \int_0^\infty \left[K_0(s)\theta^t(s) + \sum_{a=1}^{k} K_a(s)\Upsilon_a{}^t(s)\right] ds, \tag{1.35}$$

allowing $\mathfrak{F}$ to be affected by the entire history of the thermokinetic process to which $\mathcal{B}$ has been subject. In this way a theory of thermodynamic relaxation and hysteresis could be constructed.

If we review the classical treatments, we see that they did not distinguish between the general laws of thermodynamics and the constitutive equations defining particular thermodynamic bodies. The presentation I have just given you makes entirely plain not only this distinction but also the highly special classes of bodies to which the older authors confined their studies from the beginning.

Looking back at the table of primitives and axioms, some will require that the definitions in a physical theory be "operational" and that the axioms be "proved" by experiments. It is a waste of time to argue with these people, because their definitions of "operational" and "proved" are circular and hence their contentions are irrefutable (as well as insupportable). It is easiest simply to admit that they are right and leave the matter of constructing the operational definitions and the experimental proofs to them. Clearly such metaphysical paraphernalia are not a part of the *mathematical theory*, and it is only the theory that I have attempted to present. Just as clearly, such paraphernalia are equally relevant or equally irrelevant to all theories. If they are needed in thermodynamics, they are needed also in mechanics. In mechanics, questions about what force and mass really are and how they are to be measured are important, but they are rarely mentioned in books on the mathematical theory. Persons who wish to study them are referred to works on metaphysics, the philosophical-empirical foundations of physics, or experimental procedures, according to taste. With thermodynamics it is a different matter. Whenever anyone tries to do anything rational there, no matter what branch or application he takes up, he may be sure to be pestered by professionals who rise and demand he explain how to define temperature and entropy for systems not infinitesimally near to equilibrium. The persecutor himself, obviously, regards "equilibrium" and "infinitesimally near" as primitive concepts. In the theory I have presented, temperature is a primitive concept; equilibrium (of a single body only) is defined, and "near equilibrium" could be defined but is not needed except as the basis of an as yet undeveloped systematic process of approximation.

For the slightly disequilibrated professional I will make only two remarks in passing. First, in mechanics the concept of force originated in statics and was carried over bodily, if with much delay and discussion, to motions. If the restoring force exerted by a spring is proportional to the increase of its length in a static experiment, will it still be so when a ball is attached to the end and set into oscillation, especially if the experiment is performed in a spaceship in orbit around the moon? Indeed, does it make sense to talk about forces at all in a moving system? The forces, it seems, might be affected by the motions, yet we are supposed to know the forces first in order to determine what the motion will be. These questions, and far subtler ones of the same kind, were asked in the seventeenth century; today the freshman is trained specifically not to ask them. Perhaps he ought to; but if he does, he is more likely to take up philosophy than science; and if NEWTON had insisted that forces be used only when they can be measured by an operationalist who works and thinks very, very slowly, it is unlikely anyone would be designing spaceships today. The professionals who insist that temperature is a concept for equilibrium alone are not solving any problems themselves; they are merely pronouncing our problems insoluble and sneering at us for trying to solve them.

Second, the professional theorists of thermodynamics show a lack of respect† for experiment I find hard to admire. Experiments are being done

† A part of the disdain of these men must be forgiven and ignored as being merely social. They have spent their lives endeavoring, all too often with scant success, to apply to engineering problems of today some of the fundamental theories of mathematical physics set up one or two centuries ago, or by effecting an infinitesimal perturbation of some such theory to make some finite advance. In scorning those who search the fundamentals today, in belittling them for not spending their lives in slavery to structures and satellites and crystalline sandwiches, these engineers forget that the great and wide applications of the theories they themselves exalt were shown, not by the discoverers of those theories, but by generations of followers who patiently took the risk of learning first in the hope of applying later. Then, too, there were scoffers at mathematics and fundamental theory; while they sat high in the priesthoods of science and engineering, their names are forgotten, most of all by those who now squirm uneasily on their thrones.

today in all sorts of extreme conditions. My friends and colleagues tell me they measure the temperatures in polymer solutions undergoing dramatic normal-stress effects, they control the temperatures in explosion fronts, they infer the temperatures on the skins of artificial satellites. It would be presumptuous on my part to question the details of the work of these men, and I see no reason to deny that they know their business. The temperatures occurring as values of the temperature function $\theta(\cdot)$ in modern theories of materials are intended to represent the numbers these experimenters report and call "temperatures", although their systems bear little likeness to those the old books on thermodynamics describe as being "in equilibrium", and their processes seem to have little in common with whatever it is the thermodynamicists mean by "quasi-static"†. I do not claim that we yet know whether or not a particular theory be borne out by any particular experiment, but I do not see why the professionals should forbid us from trying to explain the experiments by a rational theory, approached in that same spirit everyone today regards as the right one for mechanics and electromagnetism.

There remains one aspect of classical thermodynamics not touched by this lecture so far. This aspect is the qualitative properties of irreversible processes as described by the pioneers of the last century in terms of inequalities restricting end states. One major assertion, if it can be proved, suffices as a basis for easy proofs of the rest:

Among all processes terminating at a given state, that corresponding to perpetual equilibrium in the past yields the least free energy.

The classical discussions do not prove this claim except in extremely special circumstances and at that by arguments of uncertain force. For sufficiently simple theories, *e.g.*, if $\mathfrak{F} = \Psi(\theta, \boldsymbol{\Upsilon})$ as in Examples 1 and 2, it is trivial and empty. More generally, since it refers to a limit as $t \to -\infty$, obviously it *cannot* be proved unless suitable assumptions are made regarding the limit properties of the functionals in the constitutive axiom and the nature of equilibrium. To derive a theorem of this kind, we need to solve the ***second problem of thermodynamics:*** *for an assigned class of thermodynamic bodies, to derive the principles of thermostatics.*

For this deeper aspect of thermodynamics, there is no point in remaining with the theory of homogeneous processes. The purpose of this lecture is to provide a basis for easy and natural generalization to deformable bodies. These furnish the subject of the next two lectures.

Bibliography

Some further motivation for the theory presented here may be found in another lecture:

† "A change so small that the system may be considered in equilibrium at all times" (*sic!*).

"Thermodynamics for Beginners," *Proceedings of the Symposium on Thermodynamics* (1966), Springer-Verlag, Vienna, 1968.

As mentioned there, this theory is in essence no more than a specialization of more general ideas proposed by COLEMAN & NOLL in 1963. The presentation has been influenced also by a work of WANG & BOWEN, also of a much more general nature. References to these and other sources are given in the course of the third lecture.

Lecture 2

INTERNAL DISSIPATION

Thermodynamics seems always to rest upon some prior concepts of mechanics, especially the notion of mechanical power. While the older studies leave the mechanical context unspecified, perhaps so as to lend a specious generality to the discourse, modern theorists prefer to state the mechanical basis in detail. What they thereby lose in breadth they more than gain in concreteness.

The more general the system considered, the greater is the number of variables needed to describe it. These variables are related by additional equations of balance, and to obtain a determinate theory, additional constitutive equations are needed. Greater possible diversity of response goes hand in hand with greater variety of physical circumstances. In this lecture I shall let "mechanics" mean the ordinary theory of deformable bodies. For the region of experience described by that theory, I shall construct a substructure for thermodynamics by using the concepts introduced in the preceding lecture. In the next lecture, I shall show how a thermodynamic theory formally similar to the classical one but vastly more general in scope may be set upon this substructure with the aid of suitable further axioms. The lectures following it will extend the concept of mechanics so as to allow for diffusion and chemical reactions as well.

A body $\mathcal{B}$ occupies a domain of space at each time t. These domains are its *configurations*. The mapping of one such domain or *reference configuration* $\boldsymbol{\kappa}$ onto the present configuration is called the *motion* of $\mathcal{B}$ with respect to that configuration:

$$\mathbf{x} = \mathbf{x}_{\kappa}(\mathbf{X}, t), \qquad -\infty < t < \infty. \tag{2.1}$$

Here $\mathbf{x}_{\kappa}$ is the mapping that carries the place $\mathbf{X}$ in the reference configuration $\boldsymbol{\kappa}$ into the place $\mathbf{x}$ at time t. The places $\mathbf{X}$ identify the particles, so we may speak of "the particle $\mathbf{X}$" without fear of confusion. The *velocity* $\dot{\mathbf{x}}$ is the rate of change of position of a particle:

$$\dot{\mathbf{x}} \equiv \partial_t \mathbf{x}_{\kappa}(\mathbf{X}, t) = \dot{\mathbf{x}}(\mathbf{x}, t). \tag{2.2}$$

A *mass*, which is a non-negative measure M, is defined once and for all over measurable subsets of the body and is also assumed to be an absolutely continuous function of volume in space. Hence a non-negative mass-density $\rho(\cdot)$ exists almost everywhere at each time, and at points where it and the velocity field $\dot{\mathbf{x}}(\cdot)$ are sufficiently smooth

$$\dot{\rho} + \rho \operatorname{div} \dot{\mathbf{x}} = 0. \tag{2.3}$$

The *linear momentum* $\mathfrak{m}$ and *rotational momentum* $\mathfrak{M}_{\mathbf{x}_0}$ of $\mathcal{B}$ in its present configuration are defined by

$$\mathfrak{m} \equiv \int_{\mathcal{B}} \dot{\mathbf{x}}\, dM, \qquad \mathfrak{M}_{\mathbf{x}_0} \equiv \int_{\mathcal{B}} (\mathbf{x} - \mathbf{x}_0) \wedge \dot{\mathbf{x}}\, dM, \tag{2.4}$$

where $\mathbf{x}_0$ is a fixed place and where the integration is based on mass. *Euler's laws* of mechanics assert that for every body the time rates of change $\dot{\mathfrak{m}}$ and $\dot{\mathfrak{M}}_{\mathbf{x}_0}$ of these quantities are equal to the *total force* $\mathfrak{f}$ and *total torque* $\mathfrak{F}_{\mathbf{x}_0}$ acting upon that body in its present configuration:

$$\dot{\mathfrak{m}} = \mathfrak{f}, \qquad \dot{\mathfrak{M}}_{\mathbf{x}_0} = \mathfrak{F}_{\mathbf{x}_0}. \tag{2.5}$$

In the simplest kind of continuum mechanics, the force $\mathfrak{f}$ is assumed to be the sum of two forces of special kinds: the *total body force* $\mathfrak{f}_b$ and the *total contact force* $\mathfrak{f}_c$. The total body force is an absolutely continuous function of mass, while the total contact force is an absolutely continuous function of surface area:

$$\begin{aligned} \mathfrak{f} &= \mathfrak{f}_c + \mathfrak{f}_b, \\ &= \int_{\partial\mathcal{B}} \mathbf{t}\, dA + \int_{\mathcal{B}} \mathbf{b}\, dM. \end{aligned} \tag{2.6}$$

The field $\mathbf{b}$, defined in the interior of $\mathcal{B}$, is the *body force* per unit mass; it represents forces acting at a distance, *e.g.*, those of gravitation and electromagnetism. The *traction* $\mathbf{t}$ represents the action of neighboring parts of material upon one another, or any forces applied in any way to the boundary $\partial\mathcal{B}$ of the body $\mathcal{B}$ in its present configuration. According to a theorem of CAUCHY, the traction is delivered by the *stress tensor* $\mathbf{T}$:

$$\mathbf{t} = \mathbf{T}\mathbf{n}, \tag{2.7}$$

where $\mathbf{n}$ is the outer unit normal to $\partial\mathcal{B}$. By combining $(2.4)_1$, $(2.5)_1$, (2.6), and (2.7), on the assumption that the various fields occurring in those equations be sufficiently smooth, it is easy to derive a local equivalent to the principle of linear momentum for continua:

$$\operatorname{div} \mathbf{T} + \rho\mathbf{b} = \rho\ddot{\mathbf{x}}, \tag{2.8}$$

called *Cauchy's first law of motion.*

If we restrict attention to the case when all torques are moments of forces, so that

$$\mathfrak{F}_{\mathbf{x}_0} = \int_{\partial\mathcal{B}} (\mathbf{x} - \mathbf{x}_0) \wedge \mathbf{t}\, dA + \int_{\mathcal{B}} (\mathbf{x} - \mathbf{x}_0) \wedge \mathbf{b}\, dM, \tag{2.9}$$

then another theorem of CAUCHY asserts that the stress tensor $\mathbf{T}$ is symmetric:

$$\mathbf{T} = \mathbf{T}^{\mathrm{T}}. \tag{2.10}$$

This is *Cauchy's second law of motion.*

The *kinetic energy* $\mathfrak{K}$ of $\mathfrak{B}$ in the present configuration is defined by

$$\mathfrak{K} \equiv \tfrac{1}{2} \int_{\mathfrak{B}} \dot{x}^2 \, dM, \tag{2.11}$$

while the *power* $\mathfrak{P}$ is the rate of working of all the forces acting on $\mathfrak{B}$:

$$\mathfrak{P} \equiv \int_{\partial\mathfrak{B}} \mathbf{t} \cdot \dot{\mathbf{x}} \, dA + \int_{\mathfrak{B}} \mathbf{b} \cdot \dot{\mathbf{x}} \, dM. \tag{2.12}$$

According to a simple theorem first proved by STOKES, the net working (1.6) is given by

$$\begin{aligned} \mathfrak{W} &\equiv \mathfrak{P} - \dot{\mathfrak{K}}, \\ &= \int_{\mathfrak{B}} w \, dV, \end{aligned} \tag{2.13}$$

where

$$w = \mathbf{T} \cdot \operatorname{grad} \dot{\mathbf{x}}, \tag{2.14}$$

the dot indicating the inner product in the vector space of second-order tensors ($\mathbf{A} \cdot \mathbf{B} \equiv \operatorname{tr}(\mathbf{A}\mathbf{B}^{\mathrm{T}})$). The scalar w is the net working per unit volume; it is often called the *stress power.*

These few formulae provide a specific mechanical framework upon which a structure of thermodynamics may be raised. Here, of course, all quantities are given in terms of time-dependent spatial fields, not merely functions of time alone, so we cannot blindly apply the results of Lecture 1, but we may proceed in a parallel way. A number of things I shall say will be valid equally in more general kinds of mechanics, where a result of the form (2.13) holds with a density w which is given by an expression more complicated than (2.14), allowing, for example, for the action of couple stresses, but, so as to have a fully definite and explicit mechanical background, in this lecture I shall always refer to the simple framework just set up.

The first axiom of Lecture 1, expressing the *balance of energy,* is:

$$\dot{\mathfrak{E}} = \mathfrak{W} + \mathfrak{Q}, \tag{1.7}$$

where $\mathfrak{E}$ is the internal energy, $\mathfrak{W}$ is the net working, and $\mathfrak{Q}$ is the heating. This axiom we now set alongside Euler's laws (2.5) as basic for all theories relating forces and energies to motions. Just as we have specialized Euler's laws by the assumption (2.6) so as to obtain forms appropriate to continuum mechanics, so also we may specialize the principle of balance of energy in a corresponding way. First of all, we take the net working $\mathfrak{W}$ to be that calculated according to continuum mechanics and hence to be given by (2.13) as an absolutely continuous additive set function with density (2.14). Parallel to the assumptions (2.6) and (2.9) regarding the resultant force and resultant torque, we lay down the assumption that the heating $\mathfrak{Q}$ is the sum of two heatings of special kinds: the *body heating* $\mathfrak{Q}_b$ and the *contact heating* $\mathfrak{Q}_c$; the former being an absolutely continuous function of mass, and the latter, an absolutely continuous function of surface area:

$$\begin{aligned} \mathfrak{Q} &= \mathfrak{Q}_c + \mathfrak{Q}_b, \\ &= \int_{\partial \mathfrak{B}} q\, dA + \int_{\mathfrak{B}} s\, dM. \end{aligned} \tag{2.15}$$

The volume density s is called the *heating supply;* the surface density q, the *heating influx*. The two kinds of heating, $\mathfrak{Q}_c$ and $\mathfrak{Q}_b$, are often said to describe *conduction* and *radiation,* respectively, but no such specific connotation need be made in general. Finally, the internal energy is assumed to be an absolutely continuous additive set function:

$$\mathfrak{E} = \int_{\mathfrak{B}} \varepsilon\, dM, \tag{2.16}$$

where the density ε is called the *specific internal energy* or, for short, the *energetic.*

Under the assumption (2.16), from (1.7) and (2.15) we may derive by CAUCHY's method a result inferred earlier in a major special case by FOURIER, namely, the heating influx is delivered by the *heating-flux vector*† $\mathbf{h}$:

$$q = \mathbf{h} \cdot \mathbf{n}, \tag{2.17}$$

where $\mathbf{n}$ is the outer unit normal to $\partial\mathfrak{B}$. For sufficiently smooth fields, by (2.17), (2.16), (2.15), (2.14), and (1.7) it is easy to show by use of the divergence theorem and (2.10) and (2.8) that

$$\rho\dot{\varepsilon} = w + \operatorname{div} \mathbf{h} + \rho s. \tag{2.18}$$

This differential equation is sometimes regarded as a local statement of the "first law of thermodynamics", although in fact it is a consequence of a rather subtle sequence of field definitions along with Cauchy's laws of motion (2.8) and (2.10) as well as the general balance of energy (1.7). Major special cases of this differential equation were derived, under various special hypotheses, by FOURIER, KIRCHHOFF, and C. NEUMANN.

† The terms "influx" and "efflux" are often used but are confusing because the relation of "in" and "ef" to the direction of $\mathbf{h}$ is not unequivocal. When $\mathbf{h}$ points *out* of $\partial\mathfrak{B}$, energy is flowing *in* through $\partial\mathfrak{B}$.

In the thermodynamics of homogeneous processes the axiom of irreversibility takes the form of the Clausius-Planck inequality:

$$\theta \dot{\mathfrak{H}} \geqq \mathfrak{Q}, \tag{1.14}$$

where θ is the absolute temperature and $\mathfrak{H}$ is the calory. We must now introduce the concepts of temperature and calory in continuum mechanics. The former offers no difficulty. The temperature θ is given by a positive-valued *temperature field*,

$$\theta = \theta(\mathbf{x}, t) > 0, \tag{2.19}$$

defined over the present configuration of the body $\mathcal{B}$. As in statistical mechanics, also in continuum mechanics the reciprocal of the temperature, which may be called the *coldness* ϑ, is often more convenient for the mathematical theory:

$$\vartheta \equiv \frac{1}{\theta}. \tag{2.20}$$

It is natural to assume that the calory, like the internal energy, is an absolutely continuous function of mass:

$$\mathfrak{H} = \int_{\mathcal{B}} \eta \, dM. \tag{2.21}$$

The density η, which is usually called the *specific entropy*, I prefer to name the *caloric*. This name has the virtue of dragging a red herring across the path of those who lay down police edicts concerning when and how "entropy" is to be "defined". As they back off in horror at revival of the long-dead caloric theory of heat, we can return to our concrete mathematical theory, safe behind our double negation of authority.

Caloric is introduced so as to represent gross dissipation. While caloric may flow from place to place and thus increase or decrease locally, its total, the calory $\mathfrak{H}$, is subjected to a rule favoring increase. We may regard this vague statement as expressing the content of CLAUSIUS' postulate:

$$\dot{\mathfrak{H}} \geqq 0 \qquad \text{when } \mathfrak{Q} = 0. \tag{2.22}$$

The pioneers of thermodynamics regarded this postulate as a general axiom, not restricted to any special type of body or process, but they were unable to implement it by mathematics such as to give it a concrete form. Their various "second laws" sound more like warnings or threats than principles of a rational science†. In modern continuum mechanics,

† Thus J. R. PARTINGTON in his encyclopedic work, *An Advanced Treatise on Physical Chemistry*, vol. 1, Longmans, Green & Co., London, 1949, which is surely the most abundantly documented treatise on theoretical chemistry ever published, was able to write with much justice on p. 193, "The thermodynamics of irreversible processes is entirely qualitative and of little interest in physical chemistry."

since (2.15) holds, we may express CLAUSIUS' postulate thus: In a body $\mathcal{B}$ where conditions are so balanced that

$$\int_{\partial\mathcal{B}} q\,dA + \int_{\mathcal{B}} s\,dM = 0, \tag{2.23}$$

the calory cannot decrease:

$$\int_{\mathcal{B}} \dot{\eta}\,dM \geqq 0. \tag{2.24}$$

(A body $\mathcal{B}$ is said to be in a "calorically isolated" or "insulated" configuration if $q = 0$ on $\partial\mathcal{B}$ while $s = 0$ in the interior of $\mathcal{B}$. For such a body, of course, (2.23) holds.) While this statement expresses a certain idea about irreversibility, it does not lead to a unique generalization of the Clausius-Planck inequality (1.14) to deformable bodies. In particular, since the temperature does not appear in it, it affords no connection between the ideas of heat and temperature. In this lecture I shall motivate the principle now regarded by most of the presently productive theorists, though not by all, as being a correct dissipation inequality. To this end I shall present some simple aspects of dissipation in familiar examples.

The *internal dissipation* δ is defined as being the amount by which the increase of caloric per unit coldness exceeds the local heating:

$$\delta \equiv \theta\dot{\eta} - \frac{1}{\rho}(\operatorname{div}\mathbf{h} + \rho s). \tag{2.25}$$

While δ is defined in terms of the quantities associated with temperature and heat alone, the equation of energy balance (2.18) permits us to interpret δ also as the amount by which the increase of caloric per unit coldness exceeds the increase of energetic not produced by working:

$$\delta = \theta\dot{\eta} - \left(\dot{\varepsilon} - \frac{w}{\rho}\right). \tag{2.26}$$

If we were to transfer to the local description the ideas that motivate the Clausius-Planck inequality (1.14), we should require that this increase be non-negative:

$$\delta \geqq 0. \tag{2.27}$$

We shall name (2.27) *Planck's inequality.*† For reasons to appear later,

† *Historical note.* Just attribution, always difficult, offers a spiny problem in thermodynamics, where the literature is so vague as to require almost biblical scholarship to conjecture its meaning (if any). On the one hand, one must avoid giving the impression of arrogating for recent authors ideas that are in fact old. Thus one must mention the classical names. On the other, it would be even more arrogant to attribute to old authors views they may not have had. To some extent, CLAUSIUS and PLANCK

we shall not impose it as an axiom. Instead, in the remainder of this lecture, we shall first calculate δ for the most familiar kinds of materials and then note some consequences that follow when indeed $\delta \geqq 0$.

First, for the ideal gas of the Euler-Hadamard theory,

$$\mathbf{T} = -p\mathbf{1} \tag{2.28}$$

and also, as mentioned in the first lecture,

$$\begin{aligned} \varepsilon &= \varepsilon(v, \eta), \\ p = -\partial_v \varepsilon, &\qquad \theta = \partial_\eta \varepsilon, \end{aligned} \tag{1.1}$$

where $v \equiv 1/\rho$. Hence by (2.14) and (2.3)

$$\frac{w}{\rho} = -\frac{p}{\rho}\,\mathrm{tr}\,(\mathrm{grad}\,\dot{\mathbf{x}}) = -p\dot{v}, \tag{2.29}$$

so that by (2.26)

$$\begin{aligned} \delta &= (\partial_\eta \varepsilon)\dot{\eta} + (\partial_v \varepsilon)\dot{v} - \dot{\varepsilon}, \\ &= 0. \end{aligned} \tag{2.30}$$

Thus the internal dissipation of the ideal gas is always zero. The same result holds for the general thermo-elastic material, which includes the ideal gas as a special case.

Consider now the linearly viscous fluid of the Navier-Stokes theory. For it, a viscous stress $\mathbf{V}$ is superimposed upon the Eulerian pressure:

$$\mathbf{T} = -p\mathbf{1} + \mathbf{V}, \tag{2.31}$$

and this $\mathbf{V}$ is assumed to be a homogeneous linear isotropic function of the stretching tensor $\mathbf{D}$:

$$2\mathbf{D} \equiv \mathrm{grad}\,\dot{\mathbf{x}} + (\mathrm{grad}\,\dot{\mathbf{x}})^{\mathrm{T}}, \tag{2.32}$$

---

certainly spoke of *trends in time*, and to this extent I think it just to attach their names to (2.24) and (2.27). Nevertheless, they spoke also about comparisons of *putative equilibria*, for which (2.24) and (2.27) are meaningless. It seems to me that often they and other writers on thermodynamics (including some of the present time) do not see the difference between statics and dynamics, between a differential with respect to specifying parameters and a derivative with respect to the time. Thus I have been careful to introduce the names of CLAUSIUS and PLANCK here through *association*, certainly just, rather than by specific and hence doubtful attribution.

One thing, however, must be clear. The new thermodynamics does not introduce any new basic physical axioms (*e.g.* principles of minimum dissipation or rules of symmetry). Rather, it selects among vague old ideas those that seem fit to be made general and clear, and shows that *mathematical statement* of those ideas leads by *mathematical proof* to broad and specific new results, according to the diversity of possible materials. The imposing body of theorems it has produced is based, not on any new axiom claimed to be universal, but on *rational classification* of the variety compatible with the previously known axioms, once these latter have been made precise.

and

$$\mathbf{V} = \lambda(\operatorname{tr} \mathbf{D})\mathbf{1} + 2\mu\mathbf{D}, \tag{2.33}$$

the pressure p still being related to the internal energy by $(1.1)_2$. Hence by (2.26)

$$\begin{aligned} \rho\delta &= \operatorname{tr}(\mathbf{VD}), \\ &= \lambda(\operatorname{tr} \mathbf{D})^2 + 2\mu \operatorname{tr} \mathbf{D}^2. \end{aligned} \tag{2.34}$$

Thus the internal dissipation of the Navier-Stokes fluid is the working of the viscous stress $\mathbf{V}$. The requirement $\delta \geqq 0$ is standard in this theory. Moreover, it is imposed *identically* in $\mathbf{D}$. That is, in the theory of viscous fluids we demand that the viscous stress shall never give out work in any motion we might imagine†, but it may use up work, and generally it will. It is easy to show from (2.34) that the condition $\delta \geqq 0$ for all $\mathbf{D}$ is equivalent to the classical inequalities of DUHEM and STOKES:

$$\mu \geqq 0, \qquad 3\lambda + 2\mu \geqq 0. \tag{2.35}$$

Let us now look once more at the condition $\delta \geqq 0$ in general. By (2.26), it is equivalent to

$$\theta\dot{\eta} \geqq \dot{\varepsilon} - \frac{w}{\rho}. \tag{2.36}$$

By (2.13), a deformation in which $w = 0$ for all t is one in which all mechanical work is converted into kinetic energy. If we call such a deformation *mechanically perfect*, we may conclude from (2.36) that *in a mechanically perfect deformation*

$$\theta\dot{\eta} \geqq \dot{\varepsilon}. \tag{2.37}$$

Thus in an iso-energetic perfect deformation, the caloric cannot decrease, while in an isocaloric perfect deformation, the energetic cannot increase. If we introduce the *free energetic* ψ,

$$\psi \equiv \varepsilon - \theta\eta, \tag{2.38}$$

then by (2.26)

$$\delta = \frac{w}{\rho} - \dot{\psi} - \eta\dot{\theta}, \tag{2.39}$$

so that the condition $\delta \geqq 0$ becomes

$$\dot{\psi} \leqq \frac{w}{\rho} - \eta\dot{\theta}. \tag{2.40}$$

† The classical hydrodynamicists at this point expressly rejected the other possible way of imposing the requirement, namely, as a restriction upon the *motions* possible in a fluid with arbitrarily assigned λ and μ. Had they been "operationally" inclined, they might have refused to consider all motions they could imagine instead of just those motions that *do* occur in *experiments* on *real* fluids in hard, cruel nature.

Therefore, *in a mechanically perfect deformation*

$$\dot{\psi} \leqq -\eta\dot{\theta}. \tag{2.41}$$

In particular, *in an isothermal perfect deformation, the free energetic cannot increase.* These inequalities are consonant with the system of thermodynamics sketched by PLANCK† in 1887.

Returning to (2.40), we may integrate it from the time t_0 up to the time t and obtain

$$\psi(t) - \psi(t_0) \leqq \int_{t_0}^{t} \left(\frac{w}{\rho} - \eta\dot{\theta}\right) dt. \tag{2.42}$$

If the free energetic does not decrease, $\psi(t) - \psi(t_0) \geqq 0$, so that

$$\int_{t_0}^{t} \left(\frac{w}{\rho} - \eta\dot{\theta}\right) dt \geqq 0. \tag{2.43}$$

To proceed further here, we should need to say something specific about the constitutive equation of the material. Namely, we should have to specify the way in which the material reacts to its experience, which is represented by "processes" generalizing those introduced in Lecture 1. For some materials it is possible to prove that for all processes ending at a given temperature and configuration, the free energetic is least for the constant process. A *cyclic* process is one beginning and ending in

† To obtain the two further inequalities laid down by PLANCK, we must be somewhat more specific about the mechanical description of the body. A Piola-Kirchhoff stress $\mathbf{P}$ is defined in terms of the Cauchy stress $\mathbf{T}$ as follows:

$$\mathbf{P} \equiv \frac{\mathbf{T}(\mathbf{F}^{-1})^{\mathrm{T}}}{\rho},$$

where $\mathbf{F} \equiv \operatorname{Grad} \mathbf{x}_{\kappa}$. The *enthalpy* χ and *free enthalpy* ζ are defined in terms of $\mathbf{P}$:

$$\chi \equiv \varepsilon - \mathbf{P}\cdot\mathbf{F}, \qquad \zeta \equiv \chi - \eta\theta.$$

If $\dot{\mathbf{P}} = \mathbf{0}$,

$$\theta\delta = \theta\dot{\eta} - \dot{\chi} = -\eta\dot{\theta} - \dot{\zeta}.$$

Hence the inequality $\delta \geqq 0$ takes the forms

$$\dot{\chi} \leqq \theta\dot{\eta}, \qquad \dot{\zeta} \leqq \eta\dot{\theta}.$$

Thus

$$\dot{\chi} \leqq 0 \quad \text{if} \quad \dot{\eta} = 0,$$

while

$$\dot{\zeta} \leqq 0 \quad \text{if} \quad \dot{\theta} = 0.$$

These are the further inequalities of PLANCK, rendered specific to continuum mechanics. The condition $\dot{\mathbf{P}} = \mathbf{0}$ requires that the tractions be held constant in direction while their magnitudes are so varied as to maintain constant the contact force per unit area in the reference configuration κ.

constant conditions. In such a process, then, the free energetic cannot decrease during the non-constant parts, say in the interval (t_1, t_2). From (2.43) it follows then that *in an isothermal cyclic process, the total stress power cannot be negative:*

$$\int_{t_1}^{t_2} \frac{w}{\rho}\, dt \geqq 0. \tag{2.44}$$

This result may be interpreted loosely as a statement that work cannot be gotten out of a material by deforming it without a change of temperature. Unlike the general inequality (2.42), it does not follow for all materials with non-negative internal dissipation.

The sketch I have just presented suggests how the classical assertions about special irreversible processes might be given a rational position in a mathematical theory of materials. First, part of them follow from Planck's inequality $\delta \geqq 0$ (though that inequality has not been proved necessary to get them). Second, the remainder follow if, among all processes ending with given conditions, the free energetic is least in the constant process. A rational theory that delivers these two key properties as theorems following from more plausible and general axioms will bear out the hitherto unsupported claims of the early writers on irreversible processes.

These considerations do not refer at all to heat conduction, which in the past has often been set aside and scarcely mentioned in books on thermodynamics. In the classical theory of FOURIER, heat conduction is associated with differences of temperature. Heat is said to flow from hot to cold, just as water flows downhill. If we let $\boldsymbol{\gamma}$ be the gradient of the coldness per unit coldness:

$$\boldsymbol{\gamma} \equiv \frac{1}{\vartheta} \operatorname{grad} \vartheta = -\frac{1}{\theta} \operatorname{grad} \theta, \tag{2.45}$$

then the classical *a priori* inequality is

$$-\mathbf{h} \cdot \boldsymbol{\gamma} \geqq 0. \tag{2.46}$$

That is, $-\mathbf{h}$, which points in the direction energy is flowing in through the surface on which $\mathbf{h}$ acts, makes an acute angle with the direction of greatest increase of coldness. Without adopting FOURIER's special constitutive equation for heat flow, let us consider (2.46), which we may call *Fourier's inequality*.

Suppose *both* Planck's inequality and Fourier's inequality hold. Then

$$\rho\delta - \mathbf{h} \cdot \boldsymbol{\gamma} \geqq 0. \tag{2.47}$$

But since $\operatorname{div}(\vartheta \mathbf{h}) = \vartheta(\operatorname{div} \mathbf{h} + \mathbf{h} \cdot \boldsymbol{\gamma})$, by (2.25)

$$\vartheta(\rho\delta - \mathbf{h} \cdot \boldsymbol{\gamma}) = \rho\dot{\eta} - \operatorname{div}(\vartheta \mathbf{h}) - \rho\vartheta s. \tag{2.48}$$

Hence (2.47) is equivalent to

$$\rho\dot{\eta} - \operatorname{div}(\vartheta \mathbf{h}) - \rho\vartheta s \geqq 0. \tag{2.49}$$

Integrating this inequality over the configuration of the body $\mathcal{B}$ yields by use of the divergence theorem and (2.17)

$$\dot{\mathfrak{H}} \geqq \int_{\partial\mathcal{B}} \vartheta q \, dA + \int_{\mathcal{B}} \vartheta s \, dM. \tag{2.50}$$

This condition is called the ***Clausius-Duhem inequality,*** since DUHEM was the first to introduce the surface integral on the right-hand side so as to extend to general boundaries the original statement (2.22) of CLAUSIUS about isolated systems. If the coldness field ϑ is constant at each time, (2.50) reduces to the Clausius-Planck inequality (1.14). The term with ϑs, taking account of the growth of caloric due to heating supply, first appears, it seems, in *The Classical Field Theories* (1960)†. The Clausius-Duhem inequality, the fourth and last general axiom, is a pillar of the mathematical theory of thermodynamics. (I might have said "rational thermodynamics", since to me the two terms mean the same, but I wish to make it clear that no offense is intended to those who fear that someone else might consider them to be irrational.)

The derivation of the Clausius-Duhem inequality makes it clear that, conversely, Planck's inequality and Fourier's inequality do *not* follow from it in general. By (2.48), it is equivalent to the local inequality (2.47). A little reflection shows that Fourier's inequality, correct as it is for the case envisaged by the classical theory of heat conduction, cannot be general. By placing a sufficiently strong local source of energy at a cold spot, we certainly ought to be able to force a neighboring hot spot to grow hotter rather than colder, especially if we help the process by putting a sink of energy at the hot spot. That is, heat *can* flow from cold to hot, just as water *can* flow uphill. In view of the definition (2.25), the internal dissipation δ is influenced by the local heating, be it through the supply s or the conduction div $\mathbf{h}$. According to the Clausius-Duhem inequality in the form (2.47), the heating flux $-\mathbf{h}$ must lie on or within a certain cone whose axis points along the gradient of coldness. If there is no internal dissipation, that cone is a plane normal to the gradient. If the internal dissipation is positive, the cone is *obtuse,* having an aperture which depends upon the magnitudes of the heating flux and the gradient of coldness as well as upon the coldness itself and the internal dissipation. Therefore, in the context of a theory based on the Clausius-Duhem inequality, Planck's inequality serves to exclude certain constitutive equations. A theory that forces $-\mathbf{h}$ to lie in or on the cone is consistent with the

† FLÜGGE's *Handbuch der Physik,* vol. III, part 1, Springer-Verlag, Berlin-Göttingen-Heidelberg.

Clausius-Duhem inequality, while one that permits $-\mathbf{h}$ to lie outside the cone is not.

The argument by which we have been led to the Clausius-Duhem inequality, since it rests upon the properties of certain very special materials, is not a proof of it in general. Rather, we are following here the common, tried path of theorists: We have observed a property that summarizes a number of known and understood facts, and we ask if it will serve by itself, stripped of the restricting assumptions by which we were led to it, as an axiom on the basis of which further facts may be explained and further effects predicted. In Lectures 3 and 4 we shall see that the Clausius-Duhem inequality as it stands does indeed furnish a basis for a specific theory of irreversibility, leading to plausible new results and to correlation of other known but previously disjoint facts. In Lecture 5 we shall see that in order to describe irreversibilities of mixing and chemical reactions, the Clausius-Duhem inequality will have to be given a still more general form, in which the caloric flux consists only in part in the product of the coldness by the heating flux.

If we set the Clausius-Duhem inequality,

$$\dot{\mathfrak{H}} \geqq \int_{\partial\mathfrak{B}} \vartheta q \, dA + \int_{\mathfrak{B}} \vartheta s \, dM, \tag{2.50}$$

alongside the formula for the heating,

$$\mathfrak{Q} = \int_{\partial\mathfrak{B}} q \, dA + \int_{\mathfrak{B}} s \, dM, \tag{2.15}$$

we see that with each of the two sources of heating, represented by the densities q and s, we have associated sources, represented by the densities ϑq and ϑs, for generating a lower bound for the rate of increase of caloric. In the terms used in the theory of real functions, we state that the bounds for the two rates of increase of caloric are absolutely continuous functions of the corresponding heatings and have the same derivative, which we have called the coldness ϑ.

While I have introduced the temperature field as a primitive concept in this lecture, it is not necessary to do so if we are willing to start with the heating $\mathfrak{Q}$, the calory $\mathfrak{H}$, and some assumptions of regularity. In this way temperature in a deforming, heated body may be *defined*, if we like. In fact, as has been remarked by Gurtin & Williams†, it seems to be unnecessary to assume that the two coldnesses be the same. If we call

† M. E. Gurtin & W. Williams, "An axiomatic foundation for continuum thermodynamics," *Arch. Rational Mech. Anal.*, **26**:83–117 (1967). The reader desiring a fully mathematical presentation of modern continuum thermodynamics should consult this major memoir.

ϑ_c the *contact coldness* and ϑ_b the *body coldness*, we may replace (2.50) and (2.25) by

$$\dot{\mathfrak{H}} \geqq \int_{\partial\mathcal{B}} \vartheta_c q \, dA + \int_{\mathcal{B}} \vartheta_b s \, dM, \tag{2.51}$$

$$\begin{aligned} \vartheta_c \rho \delta &\equiv \rho \dot{\eta} - (\vartheta_c \operatorname{div} \mathbf{h} + \vartheta_b \rho s), \\ &= \rho \dot{\eta} - \vartheta_c(\rho \dot{\varepsilon} - w) - (\vartheta_b - \vartheta_c)\rho s. \end{aligned} \tag{2.52}$$

All considerations in this lecture and the next remain unaffected by this apparently greater generality.

Since circumstances in which conduction and radiation of heat occur together are as yet not much studied, there seem to be few grounds for deciding whether or not a distinction between the corresponding coldnesses, as suggested by mathematical symmetry, is really necessary in physics. In similar cases occurring in other theories in the past, when the mathematics has suggested a new distinction, often a test has been made just so as to reveal it. We may remark that distinction between ϑ_c and ϑ_b enables us to consider a homogeneous, ideal gas that does not conduct heat as being nevertheless susceptible of irreversible processes of radiation, since (2.52) yields for it

$$\delta = \left(1 - \frac{\vartheta_b}{\vartheta_c}\right) s = \left(1 - \frac{\theta_c}{\theta_b}\right) s. \tag{2.53}$$

A process of this kind is sometimes introduced, if not very frankly, to describe the absorption and dispersion of ultrasonic waves due to "molecular relaxation". When (2.53) holds in a non-conducting gas, the Clausius-Duhem inequality requires that $(\theta_b - \theta_c)\rho s \geqq 0$; that is, the body temperature at a source of heating cannot be less than the contact temperature, while at a sink, it cannot be more. In this application the contact temperature θ_c is the temperature occurring in the equation of state $p = p(\rho, \theta_c)$. GURT[redacted]d that, in contrast, the two temperatures m[redacted]ommonly considered class of heat-conducting[redacted]

In this lecture I h[redacted]euristic background for the Clausius-Duhe[redacted], we have seen how it includes as special cases two of the commonest gross irreversibilities in daily life: purely mechanical dissipation, expressed by Planck's inequality, and purely thermal dissipation, expressed by Fourier's inequality. Further, we have proved that it is not equivalent to these two irreversibilities separately but is rather a generalization of them. This is as far as we can go without introducing special, distinguishing properties of materials, as embodied in *constitutive relations*. To do so, we need only follow the procedure presented in the preceding lecture for homogeneous processes.

A constitutive relation transforms the history of an arbitrary thermokinetic process into a calorodynamic process. In continuum mechanics a *thermokinetic process* is the pair of fields

$$\mathbf{x}_\kappa(\mathbf{X}, t), \qquad \theta(\mathbf{x}, t), \tag{2.54}$$

the former being a smooth mapping defined over a region $\mathcal{V}_\kappa$ in κ and the latter a scalar field with non-negative values in the image $\mathcal{V}$ of that region in the present configuration. A *calorodynamic process* (associated with a given thermokinetic process) is a set of fields

$$\mathbf{T}(\mathbf{x}, t), \qquad \mathbf{h}(\mathbf{x}, t), \qquad \varepsilon(\mathbf{x}, t), \qquad \eta(\mathbf{x}, t), \qquad \mathbf{b}(\mathbf{x}, t), \qquad s(\mathbf{x}, t), \tag{2.55}$$

defined over the region $\mathcal{V}$, and such that the *axioms of thermodynamics*, namely (2.8), (2.10), (2.18), and (2.47), are satified. The ***constitutive axiom*** or *principle of thermodynamically compatible determinism*, may be phrased much as in Lecture 1, namely: For any given body $\mathcal{B}$ at time t, *the history of the thermokinetic process to which a neighborhood of* **X** *has been subject determines a calorodynamic process for* $\mathcal{B}$ *at* **X**.

It would be tedious to reproduce in more general form the steps we have presented in detail in Lecture 1. Clearly we may regard the functionals for **b** and s as delivered by (2.8) and (2.18), respectively; we may introduce the free energetic (2.38) and derive the ***reduced dissipation inequality.*** Indeed, by (2.39) and (2.47) this latter has the form

$$\dot{\psi} - \frac{w}{\rho} + \eta\dot{\theta} + \frac{\mathbf{h}\cdot\boldsymbol{\gamma}}{\rho} \leqq 0. \tag{2.56}$$

If we compare this with (1.19), we see that apart from our referring all conditions to densities rather than to whole bodies, the only difference lies in the last term, which takes account of contact heating. The ***effective principle of thermodynamic determinism,*** extending (1.20) and of course satisfying the principle of equipresence, now is:

[redacted] (2.57)

where the functionals $\mathbb{F}$, $\mathbb{S}$, $\mathbb{m}$, $\mathbb{h}$ are such that (2.56) holds for every process (2.54). In these relations, the argument **X** is not written, since most considerations based upon them are local and hence refer to a fixed **X**.

A still simpler form is possible. Indeed, let us set†

$$\mathbf{F} \equiv \operatorname{Grad} \mathbf{x}_\kappa. \tag{2.58}$$

† "Grad" denotes the gradient operator applied to functions of **X**, while "grad" refers to functions of **x**, the time t being held constant in each case.

Then grad $\dot{\mathbf{x}} = \dot{\mathbf{F}}\mathbf{F}^{-1}$, and hence by (2.14) we have

$$\frac{w}{\rho} - \eta\dot{\theta} = \frac{1}{\rho}\mathbf{T} \cdot \operatorname{grad} \dot{\mathbf{x}} - \eta\dot{\theta}, \\ = \boldsymbol{\tau} \cdot \dot{\boldsymbol{\lambda}} \tag{2.59}$$

if we introduce the ten-dimensional vectors $\boldsymbol{\tau}$ and $\boldsymbol{\lambda}$ as the following ordered pairs:

$$\boldsymbol{\tau} \equiv \left(\frac{\mathbf{T}(\mathbf{F}^{-1})^{\mathrm{T}}}{\rho}, -\eta\right), \\ \boldsymbol{\lambda} \equiv (\mathbf{F}, \theta), \tag{2.60}$$

the first member of $\boldsymbol{\tau}$ being essentially the Piola-Kirchhoff stress tensor. Hence by (2.39) we obtain a simple expression for the internal dissipation:

$$\delta = \boldsymbol{\tau} \cdot \dot{\boldsymbol{\lambda}} - \dot{\psi}, \tag{2.61}$$

so that the reduced dissipation inequality (2.47) assumes the form

$$\dot{\psi} - \boldsymbol{\tau} \cdot \dot{\boldsymbol{\lambda}} + \boldsymbol{\iota} \cdot \boldsymbol{\gamma} \leqq 0, \tag{2.62}$$

where

$$\boldsymbol{\iota} \equiv \frac{\mathbf{h}}{\rho}. \tag{2.63}$$

The functional relations (2.57) may be written equivalently as

$$\psi(t) = \mathbb{F}(\mathbf{x}_\kappa{}^t, \theta^t), \\ \boldsymbol{\tau}(t) = \mathbb{T}(\mathbf{x}_\kappa{}^t, \theta^t), \\ \boldsymbol{\iota}(t) = \mathbb{I}(\mathbf{x}_\kappa{}^t, \theta^t). \tag{2.64}$$

The ***first problem of thermodynamics*** may be phrased now for continua in much the same way as in Lecture 1: *In an assigned class of thermo-kinetic processes* $\mathbf{x}_\kappa(\cdot)$, $\theta(\cdot)$ *and an assigned class of functionals* $\mathbb{F}$, $\mathbb{T}$, $\mathbb{I}$, *to determine all constitutive functionals,* namely, those that satisfy (2.62) identically. The problem now has a second and deeper part as well: *to determine, among all constitutive functionals* $\mathbb{F}$, *those that yield non-negative internal dissipation, at least for certain processes, and minimum free energetic in constant processes.* To both these problems, in still more general form, we shall address ourselves in the next lecture.

What I have said today should be regarded as a selection and ordering of more or less well known older ideas. In addition to the persons already mentioned, I should name ECKART† (1940) as being in principle the main

† C. ECKART, "The thermodynamics of irreversible processes. I. The simple fluid," *Phys. Rev.*, (2) **58**:267–269 (1940).

author of this approach to the Clausius-Duhem inequality. Of course, it is the fundamental researches of COLEMAN† (1964), who first revealed the nature of internal dissipation and created the modern theory of temperature and calory in deformable bodies, that make such a selection as today's a relevant and fruitful tribute to the great thermodynamicists of the last century.

† B. D. COLEMAN, "Thermodynamics of materials with memory," *Arch. Rational Mech. Anal.*, **17**:1–46 (1964), and many later papers, some in collaboration with GURTIN, MIZEL, and others. The idea that the Clausius-Duhem inequality should be interpreted as an *identical restriction* on constitutive relations and as such should lead to proof of the existence of thermodynamic potentials in certain cases is due to B. D. COLEMAN & W. NOLL, "The thermodynamics of elastic materials with heat conduction and viscosity," *Arch. Rational Mech. Anal.*, **13**:167–178 (1963).

LECTURE 3

COLEMAN'S THEOREM

The *principle of determinism*, which is one of the main metaphysical ideas underlying all of natural science, asserts that past and present experience determines present response. In this generality and vagueness, it affirms the belief that such a thing as natural science is possible, so that we have only to look for it. In the construction of a particular branch of natural science, we set up a mathematical model in which certain particular variables are made to represent the past and present, while others represent what it is they determine.

If we let f^t denote the history of the function f up to the time t:

$$f^t(s) \equiv f(t - s), \qquad 0 \leqq s < \infty, \tag{1.11}$$

then a typical expression of the principle of determinism is

$$g(t) = \mathfrak{f}(f^t), \tag{3.1}$$

where $\mathfrak{f}$ denotes a functional, that is, a function whose arguments are functions. This equation states that if the history of f up to time t is known, the value of $g(t)$ of g at time t is determined. In the more popular words favored by philosophers, f^t is the "cause", and g is the "effect". Throughout mathematical physics various special principles of determinism, involving the histories of various specific functions, are laid down as postulated *constitutive relations* and made the basis of mathematical development. All the theories of "classical" physics, whether old or new, are of this kind.

In the preceding lecture we have set up a substructure on which a thermodynamics of continuous bodies may be built. There a "cause" is the history of a thermokinetic process: $\mathbf{x}_\kappa{}^t$, θ^t, while its "effects" are the fields of free energetic ψ, caloric η, stress $\mathbf{T}$, and heating flux $\mathbf{h}$. As shown at the end of that lecture, the effective constitutive relations of thermodynamics are

$$\begin{aligned} \psi(t) &= \mathbb{F}(\mathbf{x}_\kappa{}^t, \theta^t), \\ \boldsymbol{\tau}(t) &= \mathbb{T}(\mathbf{x}_\kappa{}^t, \theta^t), \\ \boldsymbol{\iota}(t) &= \mathbb{I}(\mathbf{x}_\kappa{}^t, \theta^t), \end{aligned} \tag{2.64}$$

where the argument $\mathbf{X}$, denoting the particle, is understood to enter both sides of these relations without being written there†, and where the

† On the right-hand sides, the unwritten argument is really all $\mathbf{Z}$ in a neighborhood of $\mathbf{X}$ rather than merely $\mathbf{X}$ itself.

reduced dissipation inequality has the form

$$\dot{\psi} - \boldsymbol{\tau} \cdot \dot{\boldsymbol{\lambda}} + \boldsymbol{\iota} \cdot \boldsymbol{\gamma} \leqq 0. \qquad (2.62)$$

We repeat here the definitions of the quantities used:

$$\boldsymbol{\tau} \equiv \left(\frac{\mathbf{T}(\mathbf{F}^{-1})^T}{\rho}, -\eta\right), \qquad (2.60)_1$$

$$\boldsymbol{\iota} \equiv \frac{\mathbf{h}}{\rho}, \qquad (2.63)$$

$$\boldsymbol{\lambda} \equiv (\mathbf{F}, \theta) = (\text{Grad}\, \mathbf{x}_\kappa, \theta), \qquad (2.58), (2.60)_2$$

$$\boldsymbol{\gamma} \equiv -\frac{1}{\theta} \text{grad}\, \theta. \qquad (2.45)$$

So as to shorten later statements, we shall call $\boldsymbol{\tau}$ the *tension* and the ordered pair $(\boldsymbol{\lambda}, \boldsymbol{\gamma})$ the *site*. (The traditional term in thermodynamics is the "state", but use of it would permit the professional thermodynamicists to interrupt here with arguments about what "defines" a state, whether a state "exists", *etc.*, *etc.*) The site at **X**, of course, is determined trivially from the thermokinetic process in a neighborhood of **X**. In order to be constitutive functionals, $\mathbb{F}$, $\mathbb{T}$, and $\mathbb{I}$ must convert the history of an arbitrary thermokinetic process into a calorodynamic process. That is, the functionals $\mathbb{F}$, $\mathbb{T}$, and $\mathbb{I}$ must be such that the reduced dissipation inequality (2.62) is satisfied *identically* in the thermokinetic processes $\mathbf{x}_\kappa(\cdot)$, $\theta(\cdot)$.

The class of possible constitutive relations (2.64) is extremely general, far too general for us to be able to expect to get from it a thermodynamic theory in the classical sense, suggested by the properties of caloric "equations of state" (Lecture 1) and nineteenth-century ideas on irreversible processes (Lecture 2). Such a thermodynamic theory ought to have the following properties:

1. The free-energetic functional $\mathbb{F}$ is a *thermodynamic potential* in the sense that it determines uniquely the tension functional $\mathbb{T}$. (By (2.61), it follows that the functional $\mathbb{D}$, whose value is the internal dissipation δ, is determined likewise by the free-energetic functional $\mathbb{F}$.)
2. The equilibrium response functions are independent of the temperature gradient and obey the classical potential relations of thermostatics (essentially (1.23)).
3. The internal dissipation determined by a thermokinetic

process ending at a site of possible equilibrium is never negative.†

4. Among all thermokinetic histories ending at a site of possible equilibrium, that corresponding to constant values for all past times has the least free energetic.

Since it was COLEMAN‡ who first proved these four assertions for an important class of materials with memory, a theorem of this kind in more general form may justly be called *Coleman's theorem*.

To derive COLEMAN's theorem, it is obviously necessary to impose restrictions on the class of thermokinetic processes allowed and on the class of functionals to be considered eligible from the start. Indeed, for many important materials results of this kind do not hold.

To see this, we need only consider a classical example, the Navier-Stokes fluid, the constitutive relations for which have been given as (1.1), (2.31), (2.32). If we add the assumption that this fluid conducts heat according to Fourier's law, we may write the full set of constitutive relations as follows:

$$\begin{aligned} \mathbf{T} &= -p\mathbf{1} + \lambda(\operatorname{tr}\mathbf{D})\mathbf{1} + 2\mu\mathbf{D}, \\ p &= -\partial_v\psi, \qquad \eta = -\partial_\theta\psi, \qquad \psi = \psi(v, \theta), \\ \mathbf{h} &= \kappa\boldsymbol{\gamma}. \end{aligned} \tag{3.2}$$

These relations obviously afford an extremely special case of (2.64). They satisfy the reduced dissipation inequality identically in $\mathbf{D}$ and $\boldsymbol{\gamma}$ if and only if

$$\mu \geqq 0, \qquad 3\lambda + 2\mu \geqq 0, \qquad \kappa \geqq 0. \tag{3.3}$$

They define one of the most useful of all theories of continua, but the material they describe obviously does not possess a thermodynamics in the sense just outlined, since its free-energetic function $\psi(\cdot)$, while it does indeed determine the caloric η, does not determine the stress $\mathbf{T}$ except when the body is at rest. Thus some additional, specializing assumption such as to *exclude* the Navier-Stokes fluid is necessary if thermodynamic properties like those embodied in the classical theorem on thermodynamic potentials, as given in Example 1 of the first lecture, are to follow.

Such a specializing assumption may be sought among the ideas

† This condition is weaker than Planck's inequality (2.27), which demands that $\delta \geqq 0$ without exception. Nevertheless, the weaker inequality suffices to prove the theorem on isothermal cyclic processes (p. 34, above), since the proof of the theorem requires the inequality (2.42) only for processes passing through possible equilibrium sites.

‡ B. D. COLEMAN, "Thermodynamics of materials with memory," *Arch. Rational Mech. Anal.*, **17**:1–46 (1964), and "On thermodynamics, strain impulses, and visco-elasticity," *ibid.*, 230–254 (1964).

connected with fading memory. In practice, the experimenter always assumes that any material he touches has fading memory, that its recollection of long past experiences fades in time so as to become, eventually, altogether negligible. Without a belief to this effect, he could never interpret a laboratory experiment, for he could not know the history of any particular specimen more than a few years back, and usually he assumes that whatever happened to a specimen hours or even minutes ago is of no importance, or, what amounts to the same thing, he subjects it to a particular history over a short period which he believes will have the effect of "wiping out" its recollection of whatever may have happened before. This idea, by itself, is not enough to form a basis for thermodynamics, since certainly the Navier-Stokes-Fourier fluid forgets all its past experiences almost immediately, yet it does not enjoy the first of the four properties we have listed as belonging to a thermodynamic theory in the classical sense. Indeed, there are many kinds of fading memory.

Fading memory was first given a concrete, mathematical meaning by COLEMAN & NOLL† in 1960, and the concept and methods were extended to thermomechanics by COLEMAN in his great pioneer papers of 1964. Other students have proposed and explored other mathematical definitions of fading memory, more general or even different in kind. All such definitions stipulate some degree of *smoothness*, both for the class of functionals considered and for the processes over which they are defined. The narrower the class of processes, the broader may be that of materials for which a thermodynamic structure exists; for example, we could make precise the nineteenth-century idea of "quasi-static" processes and by restricting attention to them derive results, essentially an extension of thermostatics, for nearly all materials. For a very narrow class of materials, such as those described by the classical "equations of state", results may be derived for a vast class of processes, as shown in Example 1 of Lecture 1.

The ideas standing behind COLEMAN's theorem have been simplified and reduced to their essentials in a major memoir by GURTIN‡, the contents of which I shall now outline. The constitutive relations selected by GURTIN are of the form

$$\begin{aligned} \psi(t) &= \mathbb{F}(\lambda^t, \gamma^t), \\ \tau(t) &= \mathbb{T}(\lambda^t, \gamma^t), \\ \iota(t) &= \mathbb{I}(\lambda^t, \gamma^t). \end{aligned} \tag{3.4}$$

† B. D. COLEMAN & W. NOLL, "An approximation theorem for functionals, with applications in continuum mechanics," *Arch. Rational Mech. Anal.*, **6**:355–370 (1960).

‡ M. GURTIN, "On the thermodynamics of materials with memory," *Arch. Rational Mech. Anal.*, **28**:40–50 (1968).

Here **X** is understood to enter both sides as a parameter, playing no part in the analysis of the present lecture. For our interpretations, we shall regard $\lambda(t)$ as being $(\mathbf{F}(t), \theta(t))$ and $\gamma(t)$ as being $-$ grad $\theta(t)/\theta(t)$; in this case (3.4) are constitutive relations for a *simple material*†; but as far as the general theory is concerned, no such identification need be made, and the results can be applied to many other cases: rods, shells, multipolar media, *etc.* It is enough that $\lambda(\cdot)$ and $\gamma(\cdot)$ be functions defined on $(-\infty, k]$ for some k or on $(-\infty, +\infty)$ and have values in given finite-dimensional vector spaces $\mathfrak{L}$ and $\mathfrak{G}$.

The domain $\mathfrak{D}$ over which the constitutive functionals are defined and have certain properties of smoothness must be specified by a suitable axiom. The elements $(\lambda(\cdot), \gamma(\cdot))$ of $\mathfrak{D}$ will be called *processes.*

The concept of local linear continuation was introduced in Lecture 1 and defined by (1.24). The axiom delimiting $\mathfrak{D}$ is:

Axiom 1 (Local Closure). *$\mathfrak{D}$ is closed under local linear continuation.*

That is, given a process $(\lambda(\cdot), \gamma(\cdot))$ in $\mathfrak{D}$, and given any vectors **l** and **g** in $\mathfrak{L}$ and $\mathfrak{G}$, respectively, at any given time t_0 in the domain of $\lambda(\cdot)$, $\gamma(\cdot)$ there exists a positive k such that $(\lambda_{\mathbf{l}}, \gamma_{\mathbf{g}})$ is also in $\mathfrak{D}$ when $-\infty < t < t_0 + k$.

This axiom is extremely restrictive. First, it forces the class of processes to be very large, for it requires that from any given site which some process may reach, infinitely many distinct processes depart. Furthermore, Axiom 1 forbids us from confining the theory to processes which are either very smooth or very wild. Indeed, suppose a particular process has been wild up to the time t; then Axiom 1 tells us that from t onward, at least for a very short time, the functionals $\mathbb{F}$, $\mathbb{T}$, and $\mathbb{I}$ must remain defined for infinitely many linear and hence extremely smooth processes. Suppose, on the contrary, that a particular process has been very smooth, even constant, up to time t. Then Axiom 1 tells us that $\mathbb{F}$, $\mathbb{T}$, and $\mathbb{I}$ must remain defined for infinitely many processes that do not even have a derivative at t. This latter fact shows us that the behavior commonly called "viscous" and exemplified by (3.2) is excluded by this axiom.

† A simple material may be defined formally as one whose response to all *homogeneous* thermokinetic processes determines its response in all processes. A homogeneous process is one in which **F** and γ are uniform over the body. The greatest possible list of independent variables then is, beyond **X**,

$$t, \quad \mathbf{x}, \quad \theta^t, \quad \mathbf{F}^t, \quad \gamma^t.$$

The principle of material frame-indifference makes it possible to prove that t and **x** cannot appear as arguments of any constitutive relation. Thus we arrive at (3.4). Of course, the principle of frame-indifference affords also further restrictions upon $\mathbb{F}$, $\mathbb{T}$, $\mathbb{I}$, but they are not needed for the general development of thermodynamics.

Let $\mathfrak{D}^*$ be the set of all histories of processes. Formally,†

$$\mathfrak{D}^* = \{\lambda^*, \gamma^*: \text{for some } t, \lambda^* = \lambda^t, \gamma^* = \gamma^t, \text{ where } (\lambda(\cdot), \gamma(\cdot)) \in \mathfrak{D}\}. \quad (3.5)$$

The elements of $\mathfrak{D}^*$ are functions defined on $[0, \infty)$, and for the t occurring in (3.5) $\lambda(t) = \lambda^t(0) = \lambda^*(0)$, *etc.* If (λ^*, γ^*) is any element of $\mathfrak{D}^*$, there exists a process $(\lambda(\cdot), \gamma(\cdot))$ such that (λ^*, γ^*) is its history up to some time t_0:

$$\lambda^{t_0} = \lambda^*, \qquad \gamma^{t_0} = \gamma^*. \quad (3.6)$$

Let $\mathbf{l}$ and $\mathbf{g}$ be any vectors in $\mathfrak{L}$ and $\mathfrak{G}$. By Axiom 1, the local linear continuation $(\lambda_{\mathbf{l}}, \gamma_{\mathbf{g}})$ is a process in some interval $(-\infty, t_0 + k)$. Also, of course,

$$\begin{aligned} \lambda_{\mathbf{l}}^{t_0} &= \lambda^*, & \gamma_{\mathbf{g}}^{t_0} &= \gamma^*, \\ \lambda_{\mathbf{l}}(t_0) &= \lambda^*(0), & \gamma_{\mathbf{g}}(t_0) &= \gamma^*(0), \end{aligned} \quad (3.7)$$

and if $t_0 < t < t_0 + k$,

$$\dot{\lambda}_{\mathbf{l}}(t) = \mathbf{l}, \qquad \dot{\gamma}_{\mathbf{g}}(t) = \mathbf{g}. \quad (3.8)$$

Axiom 2 (Continuity from the Right). *In any process, the functionals* $\mathbb{F}$, $\mathbb{T}$, *and* $\mathbb{I}$ *deliver functions of t that are continuous from the right.*

This axiom asserts, roughly, that $\mathbb{F}$, $\mathbb{T}$, and $\mathbb{I}$ must describe a material that responds continuously when the body is subjected to a weak discontinuity. By Axiom 1, these functionals must be defined for certain processes not differentiable at the time t, and they may be defined for discontinuous ones. Nevertheless, their values are functions continuous from the right. There is nothing unnatural about this requirement, which is satisfied trivially in classical thermo-elasticity, but it does fail to apply to theories in which the response depends on the present time derivatives of the process, as in the Navier-Stokes theory of fluids.

Axiom 3 (Weak Chain Rule). *The function* $\psi(t)$ *delivered by the free-energetic functional* $\mathbb{F}$ *is differentiable if the process is linear. There exist functionals* $D_\lambda\mathbb{F}$, $D_\gamma\mathbb{F}$, *and* $\mathbb{D}$, *defined on* $\mathfrak{D}^*$ *and having values that are continuous from the right in any process, such that in the local linear continuation* $\lambda_{\mathbf{l}}$, $\gamma_{\mathbf{g}}$ *from the site at* t_0 *the derivative* $\dot{\psi}(t)$ *may be calculated by a chain rule as follows:*

$$\begin{aligned} \dot{\psi}(t) &= \frac{d}{dt}\mathbb{F}(\lambda_{\mathbf{l}}^t, \gamma_{\mathbf{g}}^t), \\ &= D_\lambda\mathbb{F}(\lambda_{\mathbf{l}}^t, \gamma_{\mathbf{g}}^t)\cdot\mathbf{l} + D_\gamma\mathbb{F}(\lambda_{\mathbf{l}}^t, \gamma_{\mathbf{g}}^t)\cdot\mathbf{g} - \mathbb{D}(\lambda_{\mathbf{l}}^t, \gamma_{\mathbf{g}}^t), \end{aligned} \quad (3.9)$$

provided $t_0 < t < t_0 + k$.

† The symbols λ^*, λ^t, $\lambda_{\mathbf{l}}$ always denote functions, while, since λ is used to denote a vector in $\mathfrak{L}$, we write $\lambda(\cdot)$ for a function of time whose value is such a vector.

This axiom is somewhat weaker than the *strong chain rule*, which asserts that for every process $\mathbb{F}$ delivers a differentiable function $\psi(t)$ and that further, $\dot{\psi}(t)$ may be calculated as follows when $\dot{\lambda}(t)$ and $\dot{\gamma}(t)$ exist:

$$\dot{\psi}(t) = D_\lambda\mathbb{F}(\lambda^t, \gamma^t) \cdot \dot{\lambda}(t) + D_\gamma\mathbb{F}(\lambda^t, \gamma^t) \cdot \dot{\gamma}(t) - \mathbb{D}(\lambda^t, \gamma^t). \quad (3.10)$$

If the strong chain rule holds, and if in addition the three coefficient functionals $D_\lambda\mathbb{F}$, $D_\gamma\mathbb{F}$, and $\mathbb{D}$ deliver continuous functions, the material behaves elastically in a certain sense; namely, if $\dot{\lambda}$ and $\dot{\gamma}$ suffer jump discontinuities $[\dot{\lambda}]$ and $[\dot{\gamma}]$ at t, then the jump of $\dot{\psi}$ is a homogeneous linear combination of them:

$$[\dot{\psi}](t) = D_\lambda\mathbb{F}(\lambda^t, \gamma^t) \cdot [\dot{\lambda}](t) + D_\gamma\mathbb{F}(\lambda^t, \gamma^t) \cdot [\dot{\gamma}](t) \quad (3.11)$$

where the bracket denotes the difference of limiting values at $t + 0$ and $t - 0$. The strong chain rule (3.10) thus implies that the constitutive relations deliver linear response to weak discontinuities; in rough terms, that they are compatible with instantaneous weak singularities. The result (3.11) suggests a line of thought we shall follow in the next lecture, but it is not necessary for this one, for which the weak chain rule (3.9) suffices.

For future reference we remark that Axiom 3 implies that $\dot{\psi}(t)$ is also *continuous from the right* in any local linear continuation.

The functionals $D_\lambda\mathbb{F}$, $D_\gamma\mathbb{F}$, and $\mathbb{D}$ are defined uniquely over $\mathfrak{D}^*$. Indeed, suppose that for some argument λ^*, γ^* there exist a second set of functionals $\overline{D_\lambda\mathbb{F}}$, $\overline{D_\gamma\mathbb{F}}$, and $\overline{\mathbb{D}}$ having these properties. Then we may apply (3.9) to the history of the local linear continuation $\lambda_1{}^t$, $\gamma_g{}^t$, whose properties are described by (3.7) and (3.8). In the interval when $t_0 < t < t_0 + k$, then,

$$\begin{aligned}[D_\lambda\mathbb{F}(\lambda_1{}^t, \gamma_g{}^t) - \overline{D_\gamma\mathbb{F}}(\lambda_1{}^t, \gamma_g{}^t)] \cdot \mathbf{1} \\ + [D_\gamma\mathbb{F}(\lambda_1{}^t, \gamma_g{}^t) - \overline{D_\gamma\mathbb{F}}(\lambda_1{}^t, \gamma_g{}^t)] \cdot \mathbf{g} \\ - [\mathbb{D}(\lambda_1{}^t, \gamma_g{}^t) - \overline{\mathbb{D}}(\lambda_1{}^t, \gamma_g{}^t)] = 0. \quad (3.12)\end{aligned}$$

By Axiom 3, each quantity in square brackets delivers a function of t that is continuous from the right. As $t \to t_0 + 0$, then, from (3.12) and $(3.7)_{1,2}$ we obtain

$$\begin{aligned}[D_\lambda\mathbb{F}(\lambda^*, \gamma^*) - \overline{D_\lambda\mathbb{F}}(\lambda^*, \gamma^*)] \cdot \mathbf{1} \\ + [D_\gamma\mathbb{F}(\lambda^*, \gamma^*) - \overline{D_\gamma\mathbb{F}}(\lambda^*, \gamma^*)] \cdot \mathbf{g} \\ - [\mathbb{D}(\lambda^*, \gamma^*) - \overline{\mathbb{D}}(\lambda^*, \gamma^*)] = 0. \quad (3.13)\end{aligned}$$

Since $\mathbf{1}$ and $\mathbf{g}$ are arbitrary vectors, the three quantities in brackets must vanish.

An almost equally elementary argument† serves to prove:

Theorem 1. *In order that the functionals* $\mathbb{F}$, $\mathbb{T}$, *and* $\mathbb{I}$ *satisfy the reduced dissipation inequality in every local linear continuation, it is necessary that*

$$\begin{aligned} D_\gamma \mathbb{F} &= \mathbf{0}, \\ \mathbb{T} &= D_\lambda \mathbb{F}, \\ \mathbb{D}(\lambda^t, \gamma^t) - \mathbb{I}(\lambda^t, \gamma^t) \cdot \gamma(t) &\geqq 0, \end{aligned} \tag{3.14}$$

for every process. Conversely, if (3.14) *holds, the reduced dissipation inequality is satisfied in every process for which the strong chain rule* (3.10) *is valid.*

The converse statement is immediately verified by substituting (3.14) into (3.10) so as to calculate $\dot{\psi} - \tau \cdot \dot{\lambda} + \iota \cdot \gamma$. To prove necessity, we appeal to the local linear continuation at time t_0 as before, substitute (3.4) and (3.9) into (2.62), and thus obtain

$$\begin{aligned} [D_\lambda \mathbb{F}(\lambda_1{}^t, \gamma_g{}^t) - \mathbb{T}(\lambda_1{}^t, \gamma_g{}^t)] \cdot \mathbf{1} + D_\gamma \mathbb{F}(\lambda_1{}^t, \gamma_g{}^t) \cdot \mathbf{g} \\ - \mathbb{D}(\lambda_1{}^t, \gamma_g{}^t) + \mathbb{I}(\lambda_1{}^t, \gamma_g{}^t) \cdot \gamma(t) \leqq 0. \end{aligned} \tag{3.15}$$

Since all five functionals deliver functions continuous from the right, we may calculate the limit as $t \to t_0 + 0$ as before; the outcome is

$$\begin{aligned} [D_\lambda \mathbb{F}(\lambda^*, \gamma^*) - \mathbb{T}(\lambda^*, \gamma^*)] \cdot \mathbf{1} + D_\gamma \mathbb{F}(\lambda^*, \gamma^*) \cdot \mathbf{g} \\ - \mathbb{D}(\lambda^*, \gamma^*) + \mathbb{I}(\lambda^*, \gamma^*) \cdot \gamma^*(0) \leqq 0. \end{aligned} \tag{3.16}$$

Since this inequality must hold for arbitrary vectors **1** and **g**, (3.14) follows.

Each of the conclusions of Theorem 1 has a clear and important physical meaning. The first states that the free energetic is in a certain way insensitive to changes in the present temperature gradient $\gamma(t)$, for the term proportional to $\dot{\gamma}(t)$ (that is, to **g**) in (3.9) drops out. The second states that the free-energetic functional $\mathbb{F}$ is a *thermodynamic potential* in that it determines the tension functional. That is, if the "causal" relation for the free energetic is known, so also is that for the tension: The effects of memory, whatever they may be, on the free energetic determine all effects of memory on the stress and the caloric. In the case when the strong chain rule (3.10) holds, the results $(3.14)_{1,2}$ imply that

$$\dot{\psi}(t) = \mathbb{T}(\lambda^t, \gamma^t) \cdot \dot{\lambda}(t) - \mathbb{D}(\lambda^t, \gamma^t), \tag{3.17}$$

and hence by (2.61) that

$$\delta(t) = \mathbb{D}(\lambda^t, \gamma^t). \tag{3.18}$$

That is, in view of Theorem 1, the value of the functional $\mathbb{D}$ is the internal dissipation. According to its definition (2.25), the internal dissipation is the rate of increase of caloric not compensated by conduction and body heating. From (3.18) we see that in a simple material, the internal

† This argument is due essentially to C.-C. Wang & R. M. Bowen, "On the thermodynamics of non-linear materials with quasi-elastic response," *Arch. Rational Mech. Anal.*, **22**:79–99 (1966).

dissipation equals the rate of decrease of free energetic in a process in which the site is presently kept constant. Such a connection between δ and constant processes is just what we should expect from any thermodynamic structure.

The third result, $(3.14)_3$, seems to be identical with the starting point, the Clausius-Duhem inequality in the form (2.47), but with the essential difference that the free-energetic functional $\mathbb{F}$ now determines the internal-dissipation functional $\mathbb{D}$. In general, we cannot conclude that Planck's inequality $\delta \geqq 0$ holds in all processes. From $(3.14)_3$, the most we can prove seems to be

$$\mathbb{D}(\lambda^t, \gamma^t)\big|_{\gamma(t)=\mathbf{0}} \geqq 0; \tag{3.19}$$

that is, *at such times as a process occupies a site with vanishing temperature gradient, Planck's inequality holds.* In the special case when $\mathbb{D}(\lambda^t, \gamma^t)$ is independent of γ^t, that is,

$$\mathbb{D}(\lambda^t, \gamma^t) = \mathbb{D}(\lambda^t), \tag{3.20}$$

then (3.19) yields $\delta \geqq 0$ always: *For a material in which the internal dissipation is independent of the temperature gradient, Planck's inequality holds in all processes*†. All the older theories are of this kind. (In them,

† Here is the best place to mention WANG & BOWEN's concept of quasi-elastic response, which was a major step toward the more general concept presented here. In that theory, the functionals $\mathbb{F}$, $\mathbb{T}$, and $\mathbb{I}$ are assumed to have, for any fixed process in a certain class, smooth extensions to *functions* of the present site and the time. For example,

$$\psi(t) = F(\lambda(t), \gamma(t), t),$$

where F depends on the past history of $\lambda(\cdot)$ and $\gamma(\cdot)$, that is, on $\lambda^t(s)$ and $\gamma^t(s)$ restricted to the open interval $0 < s < \infty$, and, moreover, the function F is *continuously differentiable*. This hypothesis requires the free-energetic functional $\mathbb{F}$ to have a strongly smoothing effect, for when applied to the class of processes differing from a given one only at the present instant, and hence generally discontinuous at that instant, the value of $\mathbb{F}$ is a smooth function F. Therefore a chain rule of the form (3.10) holds, and indeed

$$D_\lambda \mathbb{F} = \partial_\lambda F, \qquad D_\gamma \mathbb{F} = \partial_\gamma F, \qquad -\mathbb{D} = \partial_t F,$$

so that by $(3.14)_1$ it follows that F is *independent* of γ. Consequently, so also is $\partial_t F$, and hence $\mathbb{D}$. Thus for quasi-elastic response Planck's inequality $\delta \geqq 0$ is valid always. Such is the case also in COLEMAN's thermodynamics of materials with fading memory as he first developed it, in 1964.

In more general theories, such as the theory of materials with fading memory for which the recollection is defined in terms of the total history (λ^t, γ^t), $(3.14)_1$ does not necessarily mean that $\mathbb{F}$ is independent of γ^t. It may mean only that $\mathbb{F}$ has the same value for all histories γ^t that differ only in their present values $\gamma(t)$. In this case, (3.19) does *not* mean that Planck's inequality holds in all processes. *Cf.* B. D. COLEMAN & M. E. GURTIN, "Equipresence and constitutive equations for rigid heat conductors,"

indeed, $\mathbb{F}$ is assumed or proved to be independent of γ^t, and hence all functionals determined by it share this property.) The more general inequality (3.19) plays an essential role in the concept of equilibrium, as we shall now see. In the following structure we shall always assume that Axioms 1, 2, 3 and the results (3.14) hold.

The first step toward a theory of equilibrium is to require that the constitutive functions remain defined for processes in which the site is held fixed. These processes are of two kinds: first, those in which the site may have varied in the past but is not allowed to do so in the future; second, those in which the site has never changed. Let us introduce a notation for the *constant continuation*† f_{t_0} of a function f from the time t_0 on:

$$f_{t_0}(t) \equiv \begin{cases} f(t) & -\infty < t < t_0, \\ f(t_0) & t_0 \leqq t < \infty, \end{cases} \tag{3.21}$$

and for the *constant function* a^c with value a:

$$a^c(u) \equiv a, \qquad -\infty < u < \infty. \tag{3.22}$$

The same symbol a^c will denote also the constant history a^{ct}.

Axiom 4 (Constant Processes). *$\mathfrak{D}$ is closed under constant continuation and contains all those constant functions that have values ever assumed by any process.*

That is, if $(\lambda(\cdot), \gamma(\cdot))$ is a process defined over $(-\infty, k]$, then for every t in $(-\infty, k]$ the functions (λ_t, γ_t) and $(\lambda(t)^c, \gamma(t)^c)$ are processes defined over $(-\infty, +\infty)$. (See Figure 3.)

This axiom serves only to enable us to specify the equilibrium response of a material. If a process occupies the site $(\lambda(t), \gamma(t))$ at time t, Axiom 4 allows us to ask these questions:

1. What happens at the particle if, after it has been subjected to the given process up to the time t, it is kept at the site $(\lambda(t), \gamma(t))$ forever afterward?
2. What would have happened at time t if the particle had been kept at the site $(\lambda(t), \gamma(t))$ forever in the past?

In particular, this axiom allows us to define *equilibrium response functions* $\hat{\psi}$ and $\hat{\tau}$ as follows:

$$\begin{aligned} \hat{\psi}(\mathsf{a}, \mathsf{b}) &\equiv \mathbb{F}(\mathsf{a}^c, \mathsf{b}^c), \\ \hat{\tau}(\mathsf{a}, \mathsf{b}) &\equiv \mathbb{T}(\mathsf{a}^c, \mathsf{b}^c), \end{aligned} \tag{3.23}$$

Z. Angew. Math. Phys., **18**:199–208 (1967), especially their discussion following their Theorem 6.

For further discussion of quasi-elastic response, see the next lecture.

† These continuations have been important in continuum mechanics ever since Coleman & Noll's first researches on fading memory.

where (**a, b**) is any site occupied by any process. However, at not all such sites need $\hat{\psi}$ and $\hat{\tau}$ be presumed to enjoy properties appropriate to equilibrium, as we shall see shortly. These functions, for a suitably restricted class of sites, are those for which we shall wish to find traditional thermostatic roles.

Axiom 5 (Relaxation of Free Energetic). *For any process* $(\lambda(\cdot), \gamma(\cdot))$ *and any time* t_0,

$$\lim_{t \to \infty} \mathbb{F}(\lambda_{t_0}{}^t, \gamma_{t_0}{}^t) = \hat{\psi}(\lambda(t_0), \gamma(t_0)). \tag{3.24}$$

That is, if we hold the particle at a fixed site long enough, the free energetic of that particle will approach the value it would have had if the particle

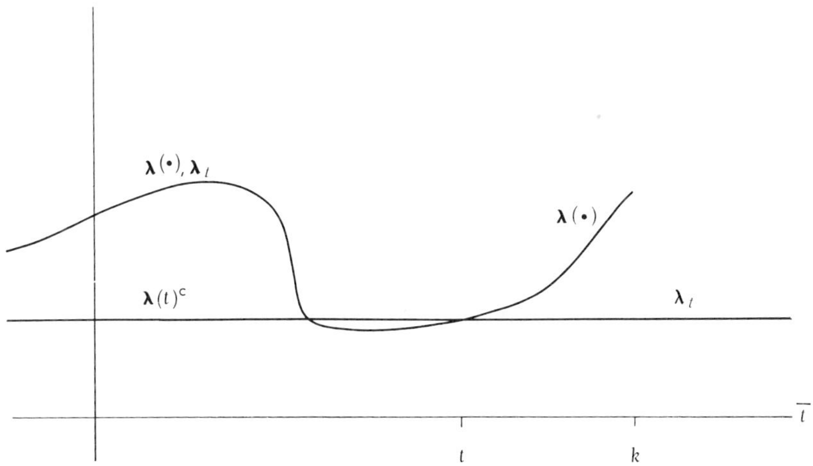

FIGURE 3. Sketch to illustrate the definitions of the functions λ_t and $\lambda(t)^c$ in terms of the function $\lambda(\cdot)$, for a fixed t. The argument of all three functions is $\bar{t}$.

had always been at that site. The two processes shown in Figure 4 are thus ultimately indistinguishable in their effects on the free energetic: In other terms, in a constant process the free energetic ultimately forgets the early experience of the material, whatever that may have been, and its response to the constant continuations of all different processes ending at the same site is ultimately the same (Figure 5). The weak kind of fading memory asserted by this axiom is nevertheless sufficient to exclude materials commonly described as "plastic", since those show rate-independent response and hence have constitutive functionals that remain unchanged under all distortions of the time axis.

We shall now use Planck's inequality, not as a general postulate or proved theorem, but as a definition of the sites possible for equilibrium.

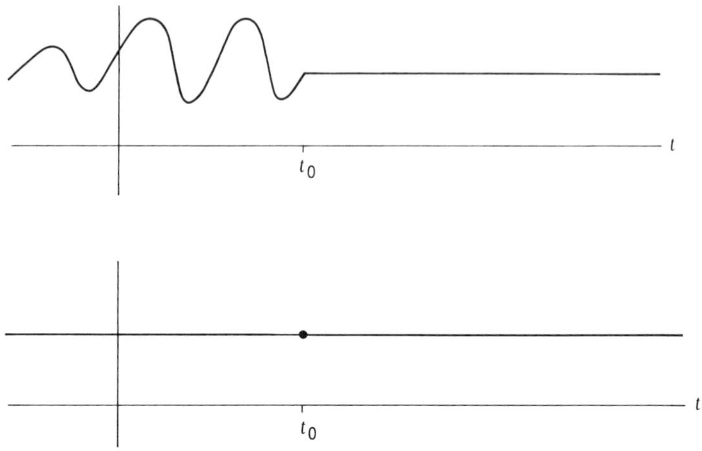

FIGURE 4. Two processes ultimately indistinguishable in their effects on the free energetic.

Let (**a**, **b**) be the site occupied by some process at some time. Then (**a**, **b**) is an *equilibrium site* if the internal dissipation is never negative in the constant continuation of any process ($\lambda(\cdot)$, $\gamma(\cdot)$) that occupies (**a**, **b**) at some time t_0:

$$\mathbb{D}(\lambda_{t_0}{}^t, \gamma_{t_0}{}^t) \geqq 0 \tag{3.25}$$

if $t > t_0$.

This definition refers to times subsequent to that at which the site is first claimed to be occupied, but we could have expressed it equally well in terms of that time itself: (**a**, **b**) *is an equilibrium site if and only if every process ending at that site at the time t produces non-negative internal dissipation then.* Indeed, by considering (3.25) in the limit as $t \to t_0 + 0$,

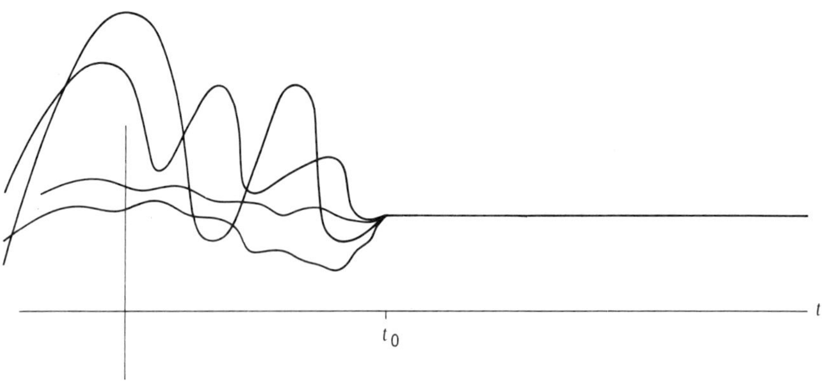

FIGURE 5. Several processes ultimately indistinguishable in their effects on the free energetic.

in virtue of the continuity from the right asserted in Axiom 3 we obtain

$$\mathbb{D}(\lambda^{t_0}, \gamma^{t_0}) \geqq 0, \tag{3.26}$$

for all processes $(\lambda(\cdot), \gamma(\cdot))$ such that $\lambda(t_0) = \mathbf{a}$, $\gamma(t_0) = \mathbf{b}$. The converse is trivial, since $(\lambda_{t_0}, \gamma_{t_0})$ is a process that occupies the site $(\lambda(t_0), \gamma(t_0))$ at the time t, if $t_0 \leqq t < \infty$.

The constant continuation $(\lambda_{t_0}, \gamma_{t_0})$ from t_0 is the linear continuation with $\mathbf{l} = \mathbf{0}$, $\mathbf{g} = \mathbf{0}$. Thus the weak chain rule (3.9) yields

$$\dot{\psi}(t) = -\mathbb{D}(\lambda_{t_0}{}^t, \lambda_{t_0}{}^t), \qquad t_0 < t < \infty. \tag{3.27}$$

Consequently, the two alternative forms of the definition of equilibrium site may be phrased equally well in terms of rate of decrease of the free energetic.

In view of the final dissipation inequality $(3.14)_3$ we may note that *if*

$$\mathbb{I}(\lambda^t, \gamma^t) \cdot \mathbf{b} = 0 \tag{3.28}$$

for all processes $(\lambda(\cdot), \gamma(\cdot))$ *occupying the site* $(\mathbf{a}, \mathbf{b})$ *at time* t, *then* $(\mathbf{a}, \mathbf{b})$ *is an equilibrium site.*

In particular, $(\mathbf{a}, \mathbf{0})$ is always an equilibrium site, provided only that there exist a process occupying it at some time. This last remark, already made, in effect, by (3.19), shows that the present concept of equilibrium site always includes the case when the temperature gradient vanishes. It may, but need not, be restricted to that case, which some students regard as the only one deserving the name "equilibrium". To leave room for some arbitrariness, we agree to consider henceforth only the equilibrium sites in a certain assigned *equilibrium space* $\mathcal{E}$, which is some connected open subset of $\mathcal{L} \times \mathcal{G}'$, where $\mathcal{G}'$ is a subspace of $\mathcal{G}$. The subspace $\mathcal{G}'$ may consist in the single vector $\gamma = \mathbf{0}$ if desired. In any case, $\mathcal{E}$ is of such a kind that it makes sense to talk about limit processes over points within it.

Axiom 6 (Smoothness of Equilibrium Response). *The equilibrium free-energetic function* $\hat{\psi}$ *is differentiable in* $\mathcal{E}$.

These axioms suffice to establish a connection between thermodynamics and thermostatics, as expressed in:

Theorem 2. *If* $(\mathbf{a}, \mathbf{b}) \in \mathcal{E}$, *then*

$$\begin{aligned} \hat{\psi}(\mathbf{a}, \mathbf{b}) &= \hat{\psi}(\mathbf{a}), \\ \hat{\tau}(\mathbf{a}, \mathbf{b}) &= \hat{\tau}(\mathbf{a}) = \partial_{\mathbf{a}}\hat{\psi}(\mathbf{a}), \end{aligned} \tag{3.29}$$

and for all processes $(\lambda(\cdot), \gamma(\cdot))$ *such that* $(\lambda(t), \gamma(t)) \in \mathcal{E}$,

$$\hat{\psi}(\lambda(t)) \leqq \mathbb{F}(\lambda^t, \gamma^t). \tag{3.30}$$

According to this theorem, the free energetic at the equilibrium site $(\mathbf{a}, \mathbf{b})$ is the same as that at $(\mathbf{a}, \mathbf{0})$. That is, the free energetic for a constant process at any site in the equilibrium space $\mathcal{E}$ is independent of what the temperature gradient there may be. (This result is empty for those particular theories in which (**a**, **b**) is not an equilibrium site *unless* $\mathbf{b} = \mathbf{0}$.) Second, the tension at equilibrium is determined from the free-energetic function as a potential by the classical formula of thermostatics. Third, among all processes ending at a given equilibrium site, that which has remained fixed at that site forever yields the least possible free energetic.

To begin the proof, let (λ^*, γ^*) be any element of $\mathfrak{D}^*$, defined by (3.5), such that $(\lambda^*(0), \gamma^*(0)) \in \mathcal{E}$. As before, there exist a time t_0 and a process $(\lambda(\cdot), \gamma(\cdot))$ such that $\lambda^{t_0} = \lambda^*$, $\gamma^{t_0} = \gamma^*$. By Axiom 4, the constant continuation $(\lambda_{t_0}, \gamma_{t_0})$ of $(\lambda(\cdot), \gamma(\cdot))$ is a process, and, since $(\lambda(t_0), \gamma(t_0)) \in \mathcal{E}$, (3.27) and (3.25) yield

$$\mathbb{F}(\lambda^*, \gamma^*) = \mathbb{F}(\lambda_{t_0}{}^{t_0}, \gamma_{t_0}{}^{t_0}) \geqq \mathbb{F}(\lambda_{t_0}{}^{t}, \gamma_{t_0}{}^{t}) \tag{3.31}$$

if $t \geqq t_0$. Now let t approach infinity. By Axiom 5 it follows that†

$$\mathbb{F}(\lambda^*, \gamma^*) \geqq \hat{\psi}(\lambda(t_0), \gamma(t_0)). \tag{3.32}$$

This is the desired result (3.30), written in a different notation, except that we have not yet shown that $\hat{\psi}$ is independent of its second argument.

To complete the proof, it is enough to show that if $(\mathbf{a}, \mathbf{b}) \in \mathcal{E}$, then

$$[\partial_{\mathbf{a}}\hat{\psi}(\mathbf{a}, \mathbf{b}) - D_{\lambda}\mathbb{F}(\mathbf{a}^c, \mathbf{b}^c)] \cdot \mathbf{l} \leqq 0 \tag{3.33}$$

for all $\mathbf{l}$ in $\mathcal{L}$, and that

$$[\partial_{\mathbf{b}}\hat{\psi}(\mathbf{a}, \mathbf{b}) - D_{\gamma}\mathbb{F}(\mathbf{a}^c, \mathbf{b}^c)] \cdot \mathbf{g} \leqq 0 \tag{3.34}$$

for all $\mathbf{g}$ in $\mathcal{G}$. Indeed, if (3.33) and (3.34) hold, the quantities in brackets must vanish, and the results $(3.14)_{1,2}$ of Theorem 1, along with the definitions (3.23), yield (3.29). We shall prove (3.33); the proof of (3.34) is parallel to it.‡

If $(\mathbf{a}, \mathbf{b}) \in \mathcal{E}$, let $(\mathbf{a}^c, \mathbf{b}^c)$ be the constant process defined by it, and at the time $t = 0$ let $(\mathbf{a}^c{}_{\mathbf{l}}, \mathbf{b}^c)$ be the linear continuation defined by $\mathbf{l}$ and $\mathbf{0}$, and let $(\mathbf{a}^c{}_{\mathbf{l}}{}^t, \mathbf{b}^c)$ denote the history of this latter up to time t. The weak chain rule (3.9) applied to this process yields

$$\frac{d}{dt}\mathbb{F}(\mathbf{a}^c{}_{\mathbf{l}}{}^t, \mathbf{b}^c) = D_{\lambda}\mathbb{F}(\mathbf{a}^c{}_{\mathbf{l}}{}^t, \mathbf{b}^c) \cdot \mathbf{l} - \mathbb{D}(\mathbf{a}^c{}_{\mathbf{l}}{}^t, \mathbf{b}^c), \tag{3.35}$$

while for the constant process $\mathbf{a}^c$, $\mathbf{b}^c$ it yields

$$\mathbb{D}(\mathbf{a}^c, \mathbf{b}^c) = 0. \tag{3.36}$$

† This part of the argument was given in essence by Coleman in his first paper on the thermodynamics of materials with memory.

‡ If $\mathcal{G}' = \{\mathbf{0}\}$, $(3.29)_1$ is trivial, so proof of (3.33) alone finishes the proof of the theorem.

By Axiom 3, $D_\lambda\mathbb{F}$ and $\mathbb{D}$ are continuous from the right, so (3.35) and (3.36) yield

$$D_\lambda\mathbb{F}(\mathbf{a}^c, \mathbf{b}^c) \cdot \mathbf{l} = \lim_{t\to 0+} \left[\frac{d}{dt}\,\mathbb{F}(\mathbf{a}^c{}_{\mathbf{l}}{}^t, \mathbf{b}^c)\right]. \tag{3.37}$$

Set

$$\Delta \equiv [\partial_{\mathbf{a}}\hat{\psi}(\mathbf{a}, \mathbf{b}) - D_\lambda\mathbb{F}(\mathbf{a}^c, \mathbf{b}^c)] \cdot \mathbf{l}. \tag{3.38}$$

By (3.37) and $(3.23)_1$,

$$\begin{aligned}\Delta &= \lim_{t\to 0+}\left[\frac{\hat{\psi}(\mathbf{a} + t\mathbf{l}, \mathbf{b}) - \hat{\psi}(\mathbf{a}, \mathbf{b})}{t} - \frac{\mathbb{F}(\mathbf{a}^c{}_{\mathbf{l}}{}^t, \mathbf{b}^c) - \mathbb{F}(\mathbf{a}^c, \mathbf{b}^c)}{t}\right],\\ &= \lim_{t\to 0+}\left[\frac{\hat{\psi}(\mathbf{a} + t\mathbf{l}, \mathbf{b}) - \mathbb{F}(\mathbf{a}^c{}_{\mathbf{l}}{}^t, \mathbf{b}^c)}{t}\right].\end{aligned} \tag{3.39}$$

Since $\mathcal{E}$ is open, for a given $\mathbf{l}$ we can find a positive t_0 such that $(\mathbf{a} + t\mathbf{l}, \mathbf{b}) \in \mathcal{E}$ when $0 \leqq t < t_0$, and we can suppose this t_0 small enough that the local linear continuation $\mathbf{a}^c{}_{\mathbf{l}}$ is linear when $0 < t < t_0$. Next, since

$$\mathbf{a}^c{}_{\mathbf{l}}{}^t(0) = \mathbf{a} + t\mathbf{l}, \qquad 0 \leqq t < t_0, \tag{3.40}$$

we conclude from (3.32) that

$$\hat{\psi}(\mathbf{a} + t\mathbf{l}, \mathbf{b}) \leqq \mathbb{F}(\mathbf{a}^c{}_{\mathbf{l}}{}^t, \mathbf{b}^c) \tag{3.41}$$

when $0 \leqq t < t_0$. Thus for all $\mathbf{l}$

$$\Delta \leqq 0, \tag{3.42}$$

so (3.39) yields (3.33), completing the proof†.

We have mentioned already that the argument given in Lecture 2 to prove the inequality (2.43), which is the basis of the theorems on cyclic processes, can be sharpened so as to apply more generally. We consider a process $(\lambda(\cdot), \gamma(\cdot))$ in which λ is continuously differentiable when $-\infty < t < t_1$, *and all whose values lie in the equilibrium space* $\mathcal{E}$. That is, roughly, all sites occupied by the process are equilibrium sites lying in the region where smooth equilibrium response is postulated. Suppose, moreover, that $(\lambda(\cdot), \gamma(\cdot))$ is a *constant* process up to time t_0, where $t_0 < t_1$, and suppose that $\lambda(t_0) = \lambda(t_1)$. That is, the material has been in equilibrium until the time t_0, and at the later time t_1 the part $\lambda(t_1)$ of the site is

† The short and elegant argument leading from (3.40) to (3.42) and allowing us to state as Axiom 6 an assumption a little weaker than its counterpart in GURTIN's paper was kindly sent me in May, 1969, by GURTIN, who attributed it to B. D. COLEMAN & D. R. OWEN.

brought back smoothly to its equilibrium value. The theorem on cyclic processes asserts that

$$\int_{t_0}^{t_1} \mathbb{T}(\lambda^t, \gamma^t) \cdot \dot{\lambda}(t)\, dt \geqq 0. \tag{3.43}$$

Roughly, the total (generalized) work done in returning the material to equilibrium is never negative: In the older terminology, a "perpetual motion of the second kind" is impossible. The proof of this fundamental result is easy from Theorems 1 and 2.

As in the case of the classical theory of "equations of state", the results here may be transformed by use of other independent variables and other potentials.

A theory such as that presented in the present lecture is mathematical. It should not be put into the hands of those who might abuse it by failure to appreciate the nature of mathematical hypothesis and proof. It is not an engineering-handbook recipe. The theorems do not refer to all materials, or to any one material in all cases. As the assumptions should make clear, a given material that obeys thermodynamic relations such as these in one class of processes may well fail to do so in others. Here, too, we have classical precedents; for example, the Navier-Stokes-Fourier fluid fits into our scheme for constant processes but not for any others.

Many particular cases are included. For example, consider the case when the vectors $\boldsymbol{\tau}$, $\boldsymbol{\iota}$, λ, γ have the standard interpretations $(2.60)_1$, (2.63), (2.58), $(2.60)_2$, and (2.45), and assume that the functionals $\mathbb{F}$, $\mathbb{T}$, and $\mathbb{I}$ obey the principle of fading memory in COLEMAN & NOLL's sense. For this case, let $\mathfrak{D}$ be the set of all histories that are continuous, are of finite recollection in the sense of a fixed obliviating measure, and have piecewise continuous derivatives which also are of finite recollection. Then a chain rule of the form (3.10) may be obtained by application of a theorem of MIZEL & WANG†, yielding the coefficient functionals as Fréchet derivatives of $\mathbb{F}$. If the equilibrium space is restricted to sites at which the temperature gradient vanishes, the relaxation property follows along the lines of the proof given in a simpler case by COLEMAN & NOLL‡.

A still simpler application may be found in the homogeneous processes of Lecture 1. There the site is given by

$$\lambda \equiv (\boldsymbol{\Upsilon}, \theta), \tag{3.44}$$

† V. J. MIZEL & C.-C. WANG, "A fading memory hypothesis which suffices for chain rules," *Arch. Rational Mech. Anal.*, **23**:124–134 (1966). *Cf.* also B. D. COLEMAN & V. J. MIZEL, "A general theory of dissipation in materials with memory," *Arch. Rational Mech. Anal.*, **27**:255–274 (1968).

‡ *E.g.*, Section 39 of *The Non-Linear Field Theories of Mechanics*, in FLÜGGE's *Handbuch der Physik*, vol. III/3, Springer-Verlag, 1965.

the vector γ being absent. If the general relations (1.20) are specialized as follows:

$$\begin{aligned} \mathfrak{W} &= -\mathbb{P}(\boldsymbol{\lambda}^t)\cdot\dot{\boldsymbol{\lambda}}, \\ \mathfrak{F} &= \mathbb{F}(\boldsymbol{\lambda}^t), \\ \mathfrak{H} &= \mathbb{H}(\boldsymbol{\lambda}^t), \end{aligned} \tag{3.45}$$

the *tension* is given by

$$\boldsymbol{\tau} \equiv (-\boldsymbol{\Pi}, -\mathfrak{H}) \tag{3.46}$$

where $\boldsymbol{\Pi} = \mathbb{P}(\boldsymbol{\lambda}^t)$. Again an assumption of fading memory suffices to derive a chain rule, *etc.*, and Theorems 1 and 2 of this lecture may be applied. It seems to have been some then unformulated but half-sensed theory of this sort that the nineteenth-century students of irreversible thermodynamics had in mind when they asserted theorems on cyclic processes. In this theory, of course, since γ is absent from the start, the final dissipation inequality $(3.14)_3$ reduces to Planck's inequality:

$$\mathbb{D}(\boldsymbol{\lambda}^t) \geqq 0, \tag{3.47}$$

that is, the internal dissipation is never negative in any process. Thus the modern general theory is consistent in all details with older theories, which took no account of the deformation of bodies or the conduction of heat.

This example serves also to remind us that the thermodynamic structure set up in this lecture is one of intermediate generality. While it vastly generalizes Example 1 of Lecture 1, it does not include Example 2, the prototype of theories of viscosity. The present theory can be enlarged to include rate effects of various kinds by postulating a chain rule (3.10) with terms proportional to $\ddot{\boldsymbol{\lambda}}$, $\dddot{\boldsymbol{\lambda}}$, . . . , $\ddot{\gamma}$, $\dddot{\gamma}$, . . . , but, as must be expected from classical examples, the results corresponding to Theorem 1 will be much more complicated. Theorem 2, since it refers only to linear processes, will hold for these more general materials without any change in statement. This, too, is obvious, since rate effects have nothing to do with equilibrium response.

COLEMAN's theorem extends to a wide class of materials with memory two classical groups of ideas: those of finite elasticity and those of thermodynamic potentials, slightly modified so as to mesh perfectly in a single, unified mathematical theory of mechanics and thermo-energetics. The realm of application is enormous. Naturally, the first trial was in the theory of sound. You will recall that elementary acoustics rests upon two ideas: first, the analysis of vibrations of a perfectly elastic continuum, which derives from NEWTON, EULER, and others, and, second, a thermodynamic theory of changes in pressure and density, lack of which held theoretical acoustics in the eighteenth century at a standstill in all questions requiring dimension-bearing answers. Eulerian hydrodynamics and the theory of caloric equations of state were brought into harmony by LAPLACE and later physicists of the nineteenth century so as to yield classical acoustics. COLEMAN's theorem extends the merger of these two ideas to the vast domain of materials with memory. In the next lecture I will show what becomes of the theory of sound in this greater range of concepts and phenomena.

Note added in March, 1969. Many people have raised doubts whether the Clausius-Duhem inequality be just as a general principle to govern irreversibility in all deformations of all materials. Such of these doubts as bubble up from centers where, as one writer has even admitted, the mathematics of the new thermodynamics is "beyond comprehension", need no rebuttal. Those who claim that the "second law" is a restriction on equilibria only, not on general motions, at bottom do no more than deny that there can be such a thing as a thermodynamics of irreversible processes. The history of physics over the centuries has seen many cries of sour grapes: "If I can't do it, it can't be done," but all such denials have been set aside by time.

Obviously, a mathematical thermodynamics of irreversible processes is possible. We have only to find it, and already we are far along the road to it. Equally obviously, the Clausius-Duhem inequality may not be the final basis, and those who are capable of using it may be capable also of modifying it or replacing it. In Lecture 5 we shall consider the proposals of I. MÜLLER. In the last lecture, I attempt a comparison with the irreversibility principle of the kinetic theory of gases. Another approach, closer to purely mechanical concepts but leading to essentially the same results as those given in the lecture above, can be based on BREUER & ONAT's concept of "recoverable work", as is shown in the major memoir by W. A. DAY, "Thermodynamics based on a work axiom," *Arch. Rational Mech. Anal.*, **31**:1–34 (1968).

LECTURE 4

WAVE PROPAGATION IN DISSIPATIVE MATERIALS

The laws of wave propagation cast light upon the nature of material response by analysis. They show how the material reacts, locally and instantaneously, to a small change or impulse in a tiny region. To compose the small effects into a motion of the body as a whole is a much harder problem even in the simplest of theories, a problem generally too hard to solve, yet we gain insight and assurance by taking the preliminary step even though we rarely follow through. The term "small" has two distinct meanings: that the disturbance itself is small, and that it is confined to a small region. There are several different approaches to wave motion, resting upon different concepts of smallness. In the commonest of these, the differential equations of motion are shorn of their non-linear terms so as to yield a linear system which may be visualized as an assembly of harmonic oscillators, whose motions may be described in the terms hallowed by centuries of contemplation of the pendulum: frequency, amplitude, wave length, phase shift. It was to this method that HUGONIOT referred when, in 1885, he wrote: " . . . hypotheses have been introduced in the equations of hydrodynamics which are disguised, it is true, under the title of *approximations*, but which singularly alter the value of such results as can be deduced from them." HUGONIOT himself developed in fairly general terms a different concept of wave propagation in which the disturbance is limited, rigorously, to a region of no volume at all, namely, a surface, but the disturbance itself may be of any amount. The resulting theory of *propagating singular surfaces* is mathematically exact. By its aid, HUGONIOT himself was the first to give fully satisfactory calculations of the speeds of sound in gases and infinitesimally elastic solids. The values of these speeds agree with those gotten earlier by the methods HUGONIOT criticized with a harshness which today would not be permitted in a young captain of naval artillery whose only other substantial publication, entitled "Study of the effects of powder in a 10-cm cannon," had appeared just the year before. Apart from rigor, the difference in concept is great. While in the method of small oscillations the surfaces of disturbance are assumed to have a special form, usually plane, and the disturbances themselves depart little from equilibrium, the method of singular surfaces allows the discontinuity to lie upon a smooth surface of any shape, and the condition of the material may be any we please. Thus, for example, the result for gases, *viz*

$$U^2 = p'(\rho), \tag{4.1}$$

while of the same *form*, has different meanings in the two cases. In the method of small oscillations, ρ is the density of the undisturbed gas, and $p'(\rho)$ is therefore a *constant:* Each part of the wave front, which is generally assumed plane, advances at the same speed in the gas otherwise *at rest*. Thus the constant U is the speed of sound in classical, *linear acoustics*. The method of singular surfaces, on the contrary, envisions an arbitrary state of gas flow, and $p'(\rho)$ is a *function* of the density, which in general varies over the flow field. Thus the function U is the speed of sound in gas dynamics, where the parts of a wave front may differ in swiftness of advance since they may pass through regions of different densities.

The method of HUGONIOT was developed by HADAMARD in his great treatise on wave motions, published in 1903, and it has always been known to specialists in mathematical gas dynamics. While it was presented in some books on physics fifty years ago, it then dropped out of the literature. With the rise of the new continuum mechanics it has come back into general use, especially in the last five years, because of its power in getting specific, concrete facts out of theories so inclusive as seemingly to defy analysis. In this lecture I shall outline the method and apply it to simple materials.

Let Ψ be a field defined and continuously differentiable in two regions $\mathcal{R}_+$ and $\mathcal{R}_-$, of which a smooth orientable surface $\mathcal{S}$ is the common boundary (Figure 6). Assume that at each point $\mathbf{x}$ on $\mathcal{S}$ the field $\Psi(\mathbf{y})$

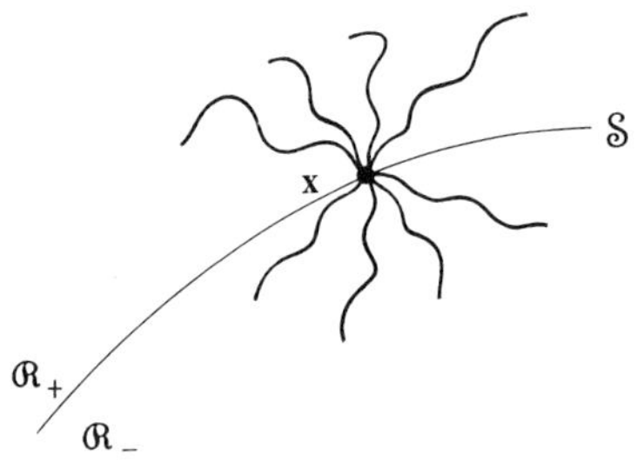

FIGURE 6. Singular surface.

approaches definite limits $\Psi^+(\mathbf{x})$ and $\Psi^-(\mathbf{x})$ as $\mathbf{y} \to \mathbf{x}$ along paths lying entirely in $\mathcal{R}_+$ and $\mathcal{R}_-$, respectively. At $\mathbf{x}$ itself the value $\Psi(\mathbf{y})$ need not be defined. The *jump* $[\Psi](\mathbf{x})$ of Ψ at $\mathbf{x}$ is the difference of the two limits at $\mathbf{x}$:

$$[\Psi] \equiv \Psi^+ - \Psi^-. \tag{4.2}$$

Assume further that the limit functions $\Psi^+(\mathbf{x})$ and $\Psi^-(\mathbf{x})$ are differentiable functions of position on the surface $\mathcal{S}$. Then $[\Psi]$ also is differentiable on $\mathcal{S}$. If $[\Psi]$ does not vanish at all points of $\mathcal{S}$, the surface $\mathcal{S}$ is said to be *singular* with respect to Ψ. A singular surface, then, is the carrier of a jump discontinuity which is spread out smoothly over it. The field Ψ may be a scalar, vector, or tensor.

The whole analysis of singular surfaces rests on a theorem stated by MAXWELL but first proved adequately by HADAMARD. Let a surface $\mathcal{S}$ be singular with respect to grad Ψ but not to Ψ itself. Then

$$[\text{grad } \Psi] = A\mathbf{n}, \tag{4.3}$$

where $\mathbf{n}$ is a vector normal to $\mathcal{S}$. That is, the jump of the gradient of a continuous field must be normal to the singular surface. In a metric space, we can choose $\mathbf{n}$ as a unit vector; in that case, A is called the *amplitude* of the singularity. The amplitude is a tensor of the same order as Ψ.

Thermomechanics allows the possibility of surfaces that are singular with respect to motion, stress, heating flux, temperature, energy, and caloric. In this lecture we shall consider only *weak* singularities, defined as those carrying jumps of the derivatives of the motion $\mathbf{x}_\kappa$ of orders 2 or greater. I shall outline the analysis only for the case when the order is exactly 2, but results of just the same form hold for higher orders.

By hypothesis, then, in the four-dimensional manifold of pairs $(\mathbf{x}, t)$, the gradient $(\mathbf{F}(\cdot), \dot{\mathbf{x}}(\cdot))$ of the motion $\mathbf{x}_\kappa$ is continuous, but the second gradient is not. The singular surface is a fixed three-dimensional surface, in the four-dimensional representation, but it is more naturally seen as the sequence of two-dimensional surfaces $\mathcal{S}$ swept out in three-dimensional space by different particles $\mathbf{X}$ of the body at different times. In general, then, $\mathcal{S}$ *propagates* through the body. If $\mathcal{S}$ is given by the equation $G(\mathbf{X}, t) = 0$ in κ, then a unit normal to it there is given by

$$\mathbf{n}_\kappa = \frac{\text{Grad } G}{|\text{Grad } G|}. \tag{4.4}$$

Consequently, if we conceive the motion of $\mathcal{S}$ as being normal to itself with speed U_κ in κ, by differentiation of $G(\mathbf{X}, t) = 0$ we obtain

$$(\text{Grad } G) \cdot U_\kappa \mathbf{n}_\kappa + G = 0; \tag{4.5}$$

by (4.4), then

$$U_\kappa = -\frac{\dot{G}}{|\text{Grad } G|}. \tag{4.6}$$

When the present configuration is chosen as the reference configuration, the corresponding speed U_κ is written simply as U and is called the *intrinsic speed of propagation*. It is the speed of normal motion of the surface $\mathcal{S}$ with respect to the particles instantaneously comprising it. If $U_\kappa \neq 0$ for some fixed κ and all t, the singular surface $\mathcal{S}$ is said to be a *wave*. It is easy to derive a relation between U and U_κ, namely,

$$U\mathbf{n}_\kappa = \mathbf{F}^T\mathbf{n}U_\kappa, \tag{4.7}$$

where $\mathbf{n}$ is the unit normal to $\mathcal{S}$ in the present configuration.

We may now apply Maxwell's theorem to the four-dimensional gradient of $\mathbf{x}_\kappa$, namely, the ordered pair $(\mathbf{F}, \dot{\mathbf{x}})$. Since the four-dimensional gradient vector to the surface $G(\mathbf{X}, t) = 0$ is $(\text{Grad}\, G, \dot{G})$, by (4.4) and (4.6) we may choose as a four-dimensional normal vector $(\mathbf{n}_\kappa, -U_\kappa)$, so that Maxwell's theorem yields *Hugoniot's conditions:*

$$\begin{aligned} [\text{Grad}\, \mathbf{F}] &= \mathbf{a} \otimes \mathbf{n}_\kappa \otimes \mathbf{n}_\kappa, \\ [\dot{\mathbf{F}}] &= -U_\kappa \mathbf{a} \otimes \mathbf{n}_\kappa, \\ [\ddot{\mathbf{x}}] &= U_\kappa{}^2 \mathbf{a}. \end{aligned} \tag{4.8}$$

Here the vector $\mathbf{a}$ is the *acceleration amplitude* of the discontinuity with respect to κ. So as to clear the writing, we do not affix a subscript κ to $\mathbf{a}$, with little danger of confusion, since change from one fixed κ to another merely multiplies $\mathbf{a}$ by a scalar, and most results of interest are independent of this change. The amplitude $\mathbf{a}$ cannot vanish, since if it did, the surface would carry no jump in any second derivative of $\mathbf{x}_\kappa$ and thus would not be a singular surface of a second order. The conditions (4.8), purely kinematical, express the hypothesis that the jump discontinuity is spread out smoothly over the surface $\mathcal{S}$ and persists in time as $\mathcal{S}$ moves. If $U \neq 0$, it is impossible that the acceleration $\ddot{\mathbf{x}}(\cdot)$ be continuous across $\mathcal{S}$. For this reason, second-order waves are often called *acceleration waves*.

To treat acceleration waves in thermodynamics, we complete their definition naturally by requiring that θ be continuous across them:

$$[\theta] = 0. \tag{4.9}$$

By Maxwell's theorem, there exists a scalar *thermal amplitude* Θ such that

$$\begin{aligned} [\text{Grad}\, \theta] &= \Theta \mathbf{n}_\kappa, \\ [\dot{\theta}] &= -U_\kappa \Theta. \end{aligned} \tag{4.10}$$

Thus an acceleration wave with given unit normal vector $\mathbf{n}_\kappa$ is characterized, at each time and at each place on $\mathcal{S}$, by the vector $\mathbf{a}$ and the scalars Θ and U_κ.

The general principles of thermomechanics, namely, the balances of linear momentum, rotational momentum, and energy, also impose restrictions upon the jumps carried by a singular surface. For a weak singular surface, balance of linear momentum is equivalent to *Poisson's condition:*

$$[\mathbf{T}]\mathbf{n} = \mathbf{0}; \tag{4.11}$$

that is, the traction is continuous, although the stress tensor need not be. Balance of rotational momentum yields the same requirement. Balance of energy is equivalent to *Fourier's condition:*

$$[\mathbf{h}] \cdot \mathbf{n} = 0. \tag{4.12}$$

That is, the contact heating is continuous, though the heating-flux vector **h** need not be. At an acceleration wave, the Clausius-Duhem inequality yields the same requirement. The proof of (4.11) and (4.12) is easy from the integral relations expressing the basic principles, providing certain assumptions of smoothness be made, and it need not be given here.

We turn now to constitutive equations. Let us first dispose of the common misconception that "dissipative mechanisms" instantly "smooth out" discontinuities. The source of this idea seems to lie in the fact that in the Navier-Stokes theory, there can be no acceleration waves. The easiest way to see how special this result is, is to derive it. The constitutive equation of the linearly viscous fluid is

$$\mathbf{T} = -p(\rho, \theta)\mathbf{1} + \lambda(\mathrm{tr}\,\mathbf{D})\mathbf{1} + 2\mu\mathbf{D}. \qquad (2.31), (2.33)$$

Hence at an acceleration wave

$$[\mathbf{T}] = \lambda[\mathrm{tr}\,\mathbf{D}]\mathbf{1} + \mu[2\mathbf{D}]. \tag{4.13}$$

Now

$$\begin{aligned} 2\mathbf{D} &\equiv \mathrm{grad}\,\dot{\mathbf{x}} + (\mathrm{grad}\,\dot{\mathbf{x}})^{\mathrm{T}}, && (2.32)\\ &= \dot{\mathbf{F}}\mathbf{F}^{-1} + (\dot{\mathbf{F}}\mathbf{F}^{-1})^{\mathrm{T}}. && (4.14) \end{aligned}$$

If we choose the present configuration as the reference configuration, as we may, then $\mathbf{F} = \mathbf{1}$, and $2\mathbf{D} = \dot{\mathbf{F}} + \dot{\mathbf{F}}^{\mathrm{T}}$, so Hugoniot's condition $(4.8)_2$ yields

$$\begin{aligned} [2\mathbf{D}] &= -U(\mathbf{a} \otimes \mathbf{n} + \mathbf{n} \otimes \mathbf{a}), \\ [\mathrm{tr}\,\mathbf{D}] &= -U\mathbf{a} \cdot \mathbf{n}, \end{aligned} \tag{4.15}$$

where **a** and **n** refer to the present configuration, and U is the intrinsic speed of propagation. Hence (4.13) becomes

$$[\mathbf{T}] = -U\{\lambda(\mathbf{a} \cdot \mathbf{n})\mathbf{1} + \mu(\mathbf{a} \otimes \mathbf{n} + \mathbf{n} \otimes \mathbf{a})\}. \tag{4.16}$$

Poisson's condition (4.11) requires, therefore, that

$$\mathbf{0} = [\mathbf{T}]\mathbf{n} = -U\{(\lambda + \mu)(\mathbf{a} \cdot \mathbf{n})\mathbf{n} + \mu\mathbf{a}\}. \tag{4.17}$$

Since $U \neq 0$ for a wave, we may cancel U from (4.17). Taking the scalar product of the result by **n** and **a** in turn yields

$$\begin{aligned} (\lambda + 2\mu)(\mathbf{a} \cdot \mathbf{n}) &= 0, \\ (\lambda + \mu)(\mathbf{a} \cdot \mathbf{n})^2 + \mu a^2 &= 0. \end{aligned} \tag{4.18}$$

As we have remarked in Lecture 2, the condition of non-negative internal dissipation requires that

$$\mu \geqq 0, \qquad 3\lambda + 2\mu \geqq 0, \tag{2.35}$$

and hence if $\mu > 0$,

$$\lambda + 2\mu \geqq \tfrac{4}{3}\mu > 0. \tag{4.19}$$

In an Eulerian fluid, $\lambda = \mu = 0$, and (4.18) reduces to $0 = 0$. In a viscous fluid, $\mu > 0$, and hence by (4.19) $\lambda + 2\mu > 0$, so (4.18) is satisfied if and only if $\mathbf{a} = \mathbf{0}$. Thus we have a classical theorem first proved by Duhem: Acceleration waves are impossible in a linearly viscous fluid. The proof rests in equal strength upon two distinct bases: the principle of dissipation and the special constitutive equation defining the Navier-Stokes theory. These together contradict Poisson's condition (4.11). For other materials, the principle of dissipation does not prohibit acceleration waves. Rather, as we shall see now, it determines their speeds of propagation and the directions of their amplitudes.

To start with, there is a broad class of dissipative materials whose constitutive equations do not contradict Poisson's condition (4.11) at an acceleration wave. In Gurtin's abstract thermodynamics as presented in the preceding lecture, we have remarked already that the basic assumptions require, more or less, that the constitutive equations be compatible with the existence of weak waves. Indeed, the local linear continuation of $(\boldsymbol{\lambda}, \boldsymbol{\gamma})$ at t produces a process $(\boldsymbol{\lambda}_l, \boldsymbol{\gamma}_g)$ which at t is continuous but has a discontinuous derivative, and we have demanded that for a local linear continuation the constitutive functionals deliver functions of t that are continuous from the right.

The constitutive equation of Fourier and Duhamel in the classical theory of heat conduction, namely,

$$\mathbf{h} = \mathbf{C}(\theta, \mathbf{F})\boldsymbol{\gamma}, \tag{4.20}$$

is enough to show us that we cannot expect the heating-flux functional $\mathbb{I}$ to remain continuous when $\boldsymbol{\gamma}$ is not. Thus in any assumptions about the smoothness of $\mathbb{I}$, special attention must be paid to its dependence upon $\boldsymbol{\gamma}$. With this much motivation, we can now strengthen Axioms 2 and 3 enough to get a concrete theory†, yet not so much as to exclude any

† The theory of waves in a broad class of materials with memory was first constructed by B. D. Coleman & M. E. Gurtin, "Waves in materials with memory, IV. Thermodynamics and the velocity of three-dimensional acceleration waves," *Arch. Rational Mech. Anal.*, **19**:317–338 (1965), reprinted in *Wave Propagation in Dissipative Materials*, Springer, New York, 1965. In this theory, as in the original theory of Coleman, the past history of the temperature gradient is not allowed as an independent variable in the constitutive equations. Coleman & Gurtin's work, which rests on the second memoir of Coleman on thermodynamics, has given rise to several generalizations, including that presented here.

classical theory other than the Navier-Stokes theory of viscous fluids:

Axiom 2⁺. *Axiom 2 holds, and moreover the value of* $\mathbb{T}$ *is a continuous function of t in any process, and the value of* $\mathbb{I}$ *is a continuous function of t in any process in which* $\gamma(t)$ *is continuous.*

Axiom 3⁺. *Axiom 3 holds, and moreover the value of* $\mathbb{D}$ *is a continuous function of t in any process.*

Not attempting to construct for waves an abstract theory at the level of GURTIN's in the preceding lecture, here we shall use Axioms 1, 2^+, and 3^+ only with the specific interpretation of $\mathbb{T}$, $\mathbb{I}$, λ, and γ expressed by (2.60), (2.63), (2.58), and (2.45). If γ is continuous at a wave, the condition $(4.10)_2$ shows that $[\dot{\theta}] = 0$ also. That is, all first derivatives of θ are continuous. Such waves are most easily pictured in the case when the temperature is uniform in space and time. For this reason, they are called *homothermal,* although in fact their definition requires only that their passage not introduce any discontinuity of the temperature or its first derivatives with respect to space and time. Thus Axiom 2^+ asserts that at such acceleration waves as correspond to processes (and by Axiom 1, there are infinitely many such), *the stress tensor and the caloric are continuous, and if the wave is homothermal, also the heating flux is continuous:*

$$[\mathbf{T}] = \mathbf{0}, \qquad [\eta] = 0, \qquad [\mathbf{h}] = \mathbf{0} \text{ if } [\boldsymbol{\gamma}] = \mathbf{0}. \tag{4.21}$$

In particular, Poisson's condition (4.11) is satisfied, and also Fourier's condition (4.12) is satisfied if the wave is homothermal. So much we assume straight off, as a restriction upon the class of constitutive equations considered. Moreover, the results of GURTIN's Theorem 1 apply *a fortiori.* Of these, for the moment we shall need only (3.18). Because of it, Axiom 3^+ imposes the requirement that *the internal dissipation be continuous:*

$$[\delta] = 0. \tag{4.22}$$

While the conditions (4.21) are assumed, in effect, as they stand, (4.22) rests upon the Clausius-Duhem inequality and hence may be regarded in part as a necessary condition for compatibility with thermodynamics.

To simplify the theory†, we assume further that the body force $\mathbf{b}$ and body heating s are continuous:

$$[\mathbf{b}] = \mathbf{0}, \qquad [s] = 0. \tag{4.23}$$

† Waves driven by discontinuous body forces and body heatings in thermoelasticity have been analyzed by M. E. GURTIN & E. K. WALSH, "Extrinsically induced acceleration waves in elastic bodies," *J. Acoust. Soc. Am.*, **41**:1320–1324 (1967). As they remark, their results could easily be generalized to materials with memory.

Each of the six conditions (4.21), (4.22), and (4.23) leads to a restriction upon other jumps, either directly through Maxwell's theorem or by use of the general principles of balance in differential form.

We shall begin with the latter kind of restriction. The Piola stress tensor $\mathbf{T}_\kappa$ is defined in terms of the Cauchy stress $\mathbf{T}$ by

$$\mathbf{T}_\kappa \equiv |\det \mathbf{F}|\mathbf{T}(\mathbf{F}^{-1})^{\mathsf{T}}. \tag{4.24}$$

It is easy to show that Cauchy's first law of motion (2.8) is equivalent to

$$\frac{1}{\rho_\kappa}\operatorname{Div}\mathbf{T}_\kappa + \mathbf{b} = \ddot{\mathbf{x}}, \tag{4.25}$$

where "Div" is the divergence operator with respect to the place $\mathbf{X}$ in $\boldsymbol{\kappa}$. Although Div $\mathbf{T}_\kappa$ and $\ddot{\mathbf{x}}$ do not exist when $\mathbf{X}$ lies on an acceleration wave, they do exist at $\mathbf{X}$ just before and just after the wave passes, and (4.25) holds then. If we assume that the constitutive functionals are such as to yield definite limits $(\operatorname{Div}\mathbf{T}_\kappa)^+$ and $(\operatorname{Div}\mathbf{T}_\kappa)^-$ as the singular surface is approached from either side along the path of $\mathbf{X}$, we obtain by $(4.23)_1$

$$\begin{aligned}\frac{1}{\rho_\kappa}(\operatorname{Div}\mathbf{T}_\kappa)^+ + \mathbf{b} &= \ddot{\mathbf{x}}^+,\\ \frac{1}{\rho_\kappa}(\operatorname{Div}\mathbf{T}_\kappa)^- + \mathbf{b} &= \ddot{\mathbf{x}}^-.\end{aligned} \tag{4.26}$$

Hence

$$\frac{1}{\rho_\kappa}[\operatorname{Div}\mathbf{T}] = [\ddot{\mathbf{x}}]. \tag{4.27}$$

Likewise, if we set

$$\mathbf{h}_\kappa \equiv |\det \mathbf{F}|\mathbf{F}^{-1}\mathbf{h}, \tag{4.28}$$

(2.27) may be written in the form

$$\frac{1}{\rho_\kappa}\operatorname{Div}\mathbf{h}_\kappa + s + \delta = \theta\dot{\eta}. \tag{4.29}$$

By (4.22), $(4.23)_2$, and the assumptions that Div $\mathbf{h}_\kappa$ approaches definite limits on each side of an acceleration wave, we obtain

$$\frac{1}{\rho_\kappa}[\operatorname{Div}\mathbf{h}_\kappa] = \theta[\dot{\eta}]. \tag{4.30}$$

Thus far, all our assumptions of smoothness have referred to the response at a fixed particle. The theory of wave motions rests also upon a *smooth variation of response* from particle to particle. Specifically, we impose an axiom on the space derivatives of the fields delivered by the constitutive equations:

Axiom 7. *Every acceleration wave is a singular surface of first order with respect to the stress field; a homothermal acceleration wave is a singular surface of first order with respect to the heating flux as well.*

In view of this axiom, we may apply Maxwell's theorem to the four-dimensional gradients of the fields $\mathbf{T}_\kappa$ and $\mathbf{h}_\kappa$. By (4.21), then,

$$\begin{aligned} U_\kappa[\mathrm{Div}\,\mathbf{T}_\kappa] &= -[\dot{\mathbf{T}}_\kappa]\mathbf{n}_\kappa, \\ U_\kappa[\mathrm{Div}\,\mathbf{h}_\kappa] &= -[\dot{\mathbf{h}}_\kappa]\cdot\mathbf{n}_\kappa \qquad \text{if} \qquad [\gamma] = \mathbf{0}. \end{aligned} \tag{4.31}$$

Eliminating the divergences between this pair of equations and (4.27) and (4.30) yields

$$\begin{aligned} [\dot{\mathbf{T}}_\kappa]\mathbf{n}_\kappa + \rho_\kappa U_\kappa[\ddot{\mathbf{x}}] &= \mathbf{0}, \\ [\dot{\mathbf{h}}_\kappa]\cdot\mathbf{n}_\kappa + \rho_\kappa \theta U_\kappa[\dot{\eta}] &= 0 \qquad \text{if} \qquad [\gamma] = \mathbf{0}. \end{aligned} \tag{4.32}$$

All space derivatives have thus been eliminated from the problem, which is reduced to determining relations among the jumps induced in certain time rates by the passage of the wave.

We recall now that $\mathbf{T}_\kappa$, η, and $\mathbf{h}_\kappa$ are delivered at $\mathbf{X}$ as functions of t by constitutive functionals whose arguments are histories of processes (λ^t, γ^t) at a given particle $\mathbf{X}$. Indeed, if we split $D_\lambda\mathbb{F}$ into components corresponding to the decomposition $\lambda = (\mathbf{F}, \theta)$:

$$D_\lambda\mathbb{F} = (D_\mathbf{F}\mathbb{F}, D_\theta\mathbb{F}), \tag{4.33}$$

then the result $(3.14)_2$ may be written in the form

$$\mathbf{T}_\kappa = \rho_\kappa D_\mathbf{F}\mathbb{F}, \qquad \eta = -D_\theta\mathbb{F}, \tag{4.34}$$

while of course the assumption $(2.64)_3$ is equivalent to

$$\mathbf{h}_\kappa = \rho_\kappa \mathbf{F}^{-1}\mathbb{I}. \tag{4.35}$$

In Lecture 3 we have seen that the strong chain rule (3.10) suffices for the validity of a jump chain rule of the form

$$[\dot{\psi}] = D_\lambda\mathbb{F}\cdot[\dot{\lambda}] + D_\gamma\mathbb{F}\cdot[\dot{\gamma}], \tag{3.11}$$

where, in fact, by $(3.14)_1$, the second term is zero. To obtain a definite theory of wave propagation, we need to have a result of this same kind for the functionals $D_\lambda\mathbb{F}$ and $\mathbb{I}$. Roughly speaking, while continuity of orders 1 and 0 for $\mathbb{F}$ and $\mathbb{I}$, respectively, was enough to derive the basic thermodynamic structure assured by Theorem 1, for the theory of waves we need continuity of orders 2 and 1, respectively. We shall express our final assumption in terms of the decomposition (4.33):

Axiom 8 (Jump chain rules). *There exist functionals* $D_{\mathbf{F},\mathbf{F}}\mathbb{F}$, $D_{\mathbf{F},\theta}\mathbb{F}$, $D_{\theta,\mathbf{F}}\mathbb{F}$, $D_{\theta,\theta}\mathbb{F}$, $D_\mathbf{F}\mathbb{I}$, $D_\theta\mathbb{I}$, and $D_\lambda\mathbb{I}$, *defined for all processes, such that at an acceleration* *wave*

$$
\begin{aligned}
\frac{1}{\rho_\kappa}[\dot{\mathbf{T}}_\kappa] &= D_{\mathbf{F},\mathbf{F}}\mathbb{F}\{[\dot{\mathbf{F}}]\} + D_{\mathbf{F},\theta}\mathbb{F}[\dot{\theta}],\\
[\dot{\eta}] &= -\operatorname{tr}(D_{\theta,\mathbf{F}}\mathbb{F}[\dot{\mathbf{F}}]^{\mathrm{T}}) - D_{\theta,\theta}\mathbb{F}[\dot{\theta}],\\
[\dot{\mathbf{h}}_\kappa] &= D_{\mathbf{F}}\mathbb{I}\{[\dot{\mathbf{F}}]\} + D_{\theta}\mathbb{I}[\dot{\theta}] + D_{\gamma}\mathbb{I}[\dot{\gamma}],
\end{aligned}
\tag{4.36}
$$

where

$$
D_{\mathbf{F},\mathbf{F}}\mathbb{F} = (D_{\mathbf{F},\mathbf{F}}\mathbb{F})^{\mathrm{T}} \qquad \text{and} \qquad D_{\mathbf{F},\theta}\mathbb{F} = D_{\theta,\mathbf{F}}\mathbb{F}. \tag{4.37}
$$

This axiom requires several comments. First, the arguments of all jumps are $\mathbf{x}$ and t, where $\mathbf{x}$ lies on the acceleration wave at time t, and the arguments of all functionals are (λ^t, γ^t). Second, the values of $D_{\mathbf{F},\mathbf{F}}\mathbb{F}$ and $D_{\mathbf{F}}\mathbb{I}$ are, respectively, a certain fourth-order tensor G and a certain third-order tensor K, and in terms of them the brace notation corresponds to the following component forms:

$$
\begin{aligned}
(\mathsf{G}\{\mathbf{A}\})_k{}^\alpha &= \mathsf{G}_k{}^\alpha{}_m{}^\beta A^m{}_\beta,\\
(\mathsf{K}\{\mathbf{A}\})^\alpha &= \mathsf{K}^\alpha{}_m{}^\beta A^m{}_\beta.
\end{aligned}
\tag{4.38}
$$

Third, the symmetry condition $(4.37)_1$ has the component form

$$
\mathsf{G}_k{}^\alpha{}_m{}^\beta = \mathsf{G}_m{}^\beta{}_k{}^\alpha. \tag{4.39}
$$

Fourth, this relation and $(4.37)_2$ express the fact that in a generalized sense the functional coefficients in $(4.36)_{1,2}$ are second partial derivatives of $\mathbb{F}$. In case $\mathbb{F}$ reduces to a function, as it does in the ordinary theory of thermo-elasticity, these relations are satisfied if the second partial derivatives of that function exist and are continuous. Fifth, that in $(4.36)_{1,2}$ no terms proportional to $[\dot{\gamma}]$ appear, is assumed because the corresponding term in the parent formula (3.11) is zero, according to $(3.14)_1$. Sixth, the jump chain rules (4.36) would follow immediately as consequences of corresponding strong chain rules for smooth processes, but since no such rules are needed for the problem at hand, we do not impose them.

While the jump chain rule axiom may seem arbitrary to the novice, those who are experienced in the modern theory of elasticity and in recent work on waves in non-linear materials will recognize it at once as containing the essence of the subject. The axiomatic structure is a little deceptive in that it does not seem to use Theorem 1 of the preceding lecture except to derive (4.22). In fact, Theorem 1 lies behind the chain rules $(4.36)_{1,2}$ in the sense that if more general formulae with terms proportional to $[\dot{\gamma}]$ were assumed, the conclusion $(3.14)_1$ of Theorem 1 would enable us to prove in many important cases that the coefficients of those terms are zero. Likewise, the conclusion $(3.14)_2$ of Theorem 1 should enable us in many cases to prove, rather than assume, the symmetry conditions (4.37).

On the other hand, even the experienced student may boggle at so many abstract axioms, one after another. He must be reminded that all these axioms state conditions of smoothness, and that at bottom the only way to get conditions of smoothness, in any theory, is to assume them as axioms. Ours here seem cumbrous only because they are *weak*. In the classical theories, stronger conditions are used regularly without even being mentioned. In thermo-elasticity, for example, all constitutive functionals reduce to functions of the present values of the thermokinetic process, and authors usually differentiate these functions freely and even invert them without so much as pointing out that axioms are needed in order to justify their operations. Of course, often one condition of smoothness can be deduced from another which may be more plausible or at least more concise. So as to realize our axioms in an example closer to

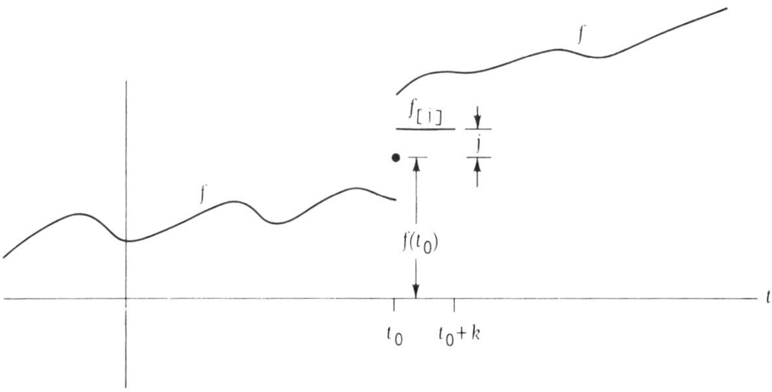

FIGURE 7. Local jump continuation.

common experience, before going on with the theory of waves I insert here a brief explanation of WANG & BOWEN's concept of quasi-elastic response†, already mentioned in the preceding lecture.

In the original work of COLEMAN, the class of histories for which the constitutive functionals were defined was presumed great enough to include *small local jump continuations*. Such a continuation $f_{[j]}$ of a function f is defined as follows (Figure 7):

$$f_{[j]} \equiv \begin{cases} f(t), & -\infty < t < t_0 \\ f(t_0) + j, & t_0 \leqq t < k. \end{cases} \tag{4.40}$$

While the material need not be assumed to respond smoothly to such a history, with the domain of definition so enlarged it makes sense to discuss the change produced in the value of a functional $\mathfrak{g}(f^t)$ when the past

† C.-C. WANG & R. M. BOWEN, "On the thermodynamics of non-linear materials with quasi-elastic response," *Arch. Rational Mech. Anal.*, **22**:79–99 (1966).

history is held fixed but the present value $f(t)$ is given a sufficiently small increment. If we use the symbol $f_+{}^t$ to denote $f^t(s)$ restricted to arguments with positive s, the functional $\mathfrak{g}(f^t)$ may now be written as

$$\mathfrak{g}(f^t) = \mathbb{h}(f_+{}^t, f(t)). \tag{4.41}$$

For any fixed past history, then, all that can vary in (4.41) is the value f of $f(t)$:

$$\mathfrak{g}(f^t) = F(f), \qquad t \text{ fixed.} \tag{4.42}$$

Thus as t varies along a given history f^t, (4.41) has a representation of the form

$$\mathfrak{g}(f^t) = F(f, t). \tag{4.43}$$

If the function F is *continuously differentiable* with respect to both arguments f and t, for the history f^t of each process in some assigned class, the original functional $\mathfrak{g}$ is said to be *quasi-elastic* for that class of processes. The reason for the name is clear from (4.42), which is the prototype of a "stress-strain reaction" in elasticity, except that here the functional relation may be also time-dependent, as indicated by (4.43).

A body *suffers quasi-elastic response* for a class of processes satisfying Axiom 1 of the preceding lecture if the response functionals $\mathbb{F}$, $\mathbb{T}$, and $\mathbb{I}$ for its particles $\mathbf{X}$ are quasi-elastic in that class and are continuously differentiable with respect to $\mathbf{X}$. We denote the corresponding functions by P, $\mathbf{T}$, and $\mathbf{i}$:

$$\begin{aligned} \psi(\mathbf{x}, t) &= P(\boldsymbol{\lambda}, \boldsymbol{\gamma}, t, \mathbf{X}), \\ \boldsymbol{\tau}(\mathbf{x}, t) &= \mathbf{T}(\boldsymbol{\lambda}, \boldsymbol{\gamma}, t, \mathbf{X}), \\ \mathbf{h}(\mathbf{x}, t) &= \rho\mathbf{i}(\boldsymbol{\lambda}, \boldsymbol{\gamma}, t, \mathbf{X}), \end{aligned} \tag{4.44}$$

where $\mathbf{x}$ is the position into which $\mathbf{X}$ is carried at time t by the motion $\mathbf{x}_\kappa$ of the body, according to (2.1). The functions P, $\mathbf{T}$, and $\mathbf{i}$, of course, are not material functions in general, since even at the same $\mathbf{X}$ a different past history $\boldsymbol{\lambda}_+{}^t$, $\boldsymbol{\gamma}_+{}^t$ will give rise to a different dependence on t and hence to *different functions* P, $\mathbf{T}$, and $\mathbf{i}$.

Axiom 2^+ is more than satisfied by the assumption of quasi-elastic response, and the strong chain rule (3.10), which is more than sufficient for Axiom 3^+, follows by differentiating P. Thus Theorem 1 of the preceding lecture holds here, and $(3.14)_{1,2}$ and (3.18) yield

$$\begin{aligned} \psi(\mathbf{x}, t) &= P(\boldsymbol{\lambda}, t, \mathbf{X}), \\ \boldsymbol{\tau}(\mathbf{x}, t) &= \partial_{\boldsymbol{\lambda}} P(\boldsymbol{\lambda}, t, \mathbf{X}), \\ \delta(\mathbf{x}, t) &= -\partial_t P(\boldsymbol{\lambda}, t, \mathbf{X}), \end{aligned} \tag{4.45}$$

where $\partial_{\boldsymbol{\lambda}}$ denotes the partial gradient with respect to $\boldsymbol{\lambda}$, and ∂_t the partial derivative with respect to t. Since the function $\mathbf{T}$ is continuously differentiable, from $(4.45)_{1,2}$ we obtain

$$\dot{\boldsymbol{\tau}} = \partial_{\boldsymbol{\lambda}}{}^2 P(\boldsymbol{\lambda}, t, \mathbf{X})\{\dot{\boldsymbol{\lambda}}\} + \partial_t \partial_{\boldsymbol{\lambda}} P(\boldsymbol{\lambda}, t, \mathbf{X}), \tag{4.46}$$

where $\partial_{\boldsymbol{\lambda}}{}^2 P$ and $\partial_t \partial_{\boldsymbol{\lambda}} P$ are continuous functions, and hence at an acceleration wave

$$[\dot{\boldsymbol{\tau}}] = \partial_{\boldsymbol{\lambda}}{}^2 P(\boldsymbol{\lambda}, t, \mathbf{X})\{[\dot{\boldsymbol{\lambda}}]\}. \tag{4.47}$$

If we split this relation into components corresponding to the decomposition $\boldsymbol{\lambda} = (\mathbf{F}, \theta)$, we obtain $(4.36)_{1,2}$, with the following specific coefficients:

$$\begin{aligned} D_{\mathbf{F},\mathbf{F}}\mathbb{F} &= \partial_{\mathbf{F}}{}^2 P, & D_{\mathbf{F},\theta}\mathbb{F} &= \partial_{\mathbf{F}}\partial_\theta P, \\ D_{\theta,\mathbf{F}}\mathbb{F} &= \partial_\theta \partial_{\mathbf{F}} P, & D_{\theta,\theta}\mathbb{F} &= \partial_\theta{}^2 P. \end{aligned} \tag{4.48}$$

Since, by hypothesis, the right-hand sides are continuous functions of $\mathbf{F}$ and θ, the symmetry relations (4.37) follow. In a similar way, $(4.36)_3$ may be derived. Thus the hypothesis of quasi-elastic response suffices for the truth of Axioms 2^+, 3^+, and 8. As for Axiom 7, the hypothesis of quasi-elastic response and the result $(4.45)_2$ yield at points not on the wave a chain rule of the form

$$\partial_{\mathbf{X}}\boldsymbol{\tau} = \rho_\kappa(\partial_{\boldsymbol{\gamma}}{}^2 P)\partial_{\mathbf{X}}\boldsymbol{\lambda} + \rho_\kappa \partial_{\mathbf{X}}\partial_{\boldsymbol{\lambda}} P + (\partial_{\mathbf{X}}\rho_\kappa)\partial_{\boldsymbol{\lambda}} P. \tag{4.49}$$

Since the last two terms and the coefficient of $\partial_{\mathbf{X}}\boldsymbol{\lambda}$ are continuous functions of $\boldsymbol{\lambda}$, t, and $\mathbf{X}$ everywhere, the fact that $\mathcal{S}$ is singular of first order with respect to $\boldsymbol{\lambda}$ makes it singular of first order with respect to $\boldsymbol{\tau}$ also. Similar analysis may be applied to the heating flux, for which, since $\boldsymbol{\gamma}$ remains an argument of $\mathbf{i}$, continuity follows only in processes such that $\boldsymbol{\gamma}(t)$ is continuous. Hence Axiom 7 is satisfied.

Ordinary thermo-elasticity furnishes a concrete special case of quasi-elastic response, obtained by dropping t as an independent variable in the functions on the right-hand side of (4.44). (In this case P, $\mathbf{T}$, and $\mathbf{i}$ are material functions.) Thus our Axioms 1, 2^+, 3^+, 7, and 8 are satisfied trivially for this classical theory. In fact, they abstract the essence of it insofar as wave propagation is concerned, for they render explicit those properties that make possible undular response of an essentially elastic kind, as we shall now see.

Turning our attention back to the general theory†, based on Axioms 1, 2^+, 3^+, 7, and 8, we first remark that if G is a fourth-order tensor, then by $(4.38)_1$ $\mathsf{G}\{\mathbf{v} \otimes \mathbf{w}\}\mathbf{w}$ is a homogeneous linear function of the vector $\mathbf{v}$ and hence is represented by the operation of a second-order tensor $\mathsf{G}(\mathbf{w})$ acting on $\mathbf{v}$:

$$\mathsf{G}\{\mathbf{v} \otimes \mathbf{w}\}\mathbf{w} = \mathsf{G}(\mathbf{w})\mathbf{v}. \tag{4.50}$$

† The elegant derivation which follows here, based upon (4.31) and (4.50) and thus avoiding the usual explicit calculation of the components of div $\mathbf{T}_\kappa$, was disclosed to me by GURTIN in a letter dated February 26, 1968.

Moreover, if $\mathbf{G}$ is symmetric in the sense (4.39), $\mathbf{G}$ is a symmetric tensor:

$$\mathbf{G} = \mathbf{G}^{\mathsf{T}}. \tag{4.51}$$

Thus, in particular, there exists a functional $\mathbb{A}(\lambda^t, \gamma^t, \mathbf{n}_\kappa)$ such that

$$D_{\mathbf{F},\mathbf{F}}\mathbb{F}\{\mathbf{a} \otimes \mathbf{n}_\kappa\}\mathbf{n}_\kappa = \mathbb{A}\mathbf{a}, \tag{4.52}$$

where for a fixed history (λ^t, γ^t) the value of $\mathbb{A}$ is a function $\tilde{\mathbf{A}}(\mathbf{n}_\kappa)$ whose values, in turn, are symmetric tensors†:

$$\mathbb{A}(\lambda^t, \gamma^t, \mathbf{n}_\kappa) = \tilde{\mathbf{A}}(\mathbf{n}_\kappa) = \tilde{\mathbf{A}}(\mathbf{n}_\kappa)^{\mathsf{T}}. \tag{4.53}$$

We now eliminate $[\dot{\mathbf{T}}]$ between $(4.32)_1$ and $(4.36)_1$, and in the result we substitute the kinematical conditions $(4.8)_{2,3}$ and $(4.10)_2$. Thus we find that

$$(\tilde{\mathbf{A}}(\mathbf{n}_\kappa) - U_\kappa{}^2\mathbf{1})\mathbf{a} + \Theta\mathbf{L}(\mathbf{n}_\kappa) = \mathbf{0}, \tag{4.54}$$

where we have used (4.52), and where

$$\mathbf{L}(\mathbf{n}_\kappa) = (D_{\mathbf{F},\theta}\mathbb{F})\mathbf{n}_\kappa. \tag{4.55}$$

Because of $(4.37)_2$, similar analysis applied to $(4.36)_2$ yields

$$[\dot{\eta}] = U_\kappa(\mathbf{L}(\mathbf{n}_\kappa) \cdot \mathbf{a} - C\Theta), \tag{4.56}$$

where

$$C \equiv -D_{\theta,\theta}\mathbb{F}. \tag{4.57}$$

There are two important cases when we need go no further in order to get definite results. First, for *homothermal waves* (4.54) reduces to a classical *propagation condition:*

$$(\tilde{\mathbf{A}}(\mathbf{n}_\kappa) - U_\kappa{}^2\mathbf{1})\mathbf{a} = \mathbf{0}. \tag{4.58}$$

In the second special case, $[\dot{\eta}] = 0$, so that, by Maxwell's theorem, $[\operatorname{Grad} \eta] = \mathbf{0}$ also. Waves of this kind, since they carry no jump of any derivative of the caloric, are called *homocaloric.* For them, by (4.56),

$$C\Theta = \mathbf{L}(\mathbf{n}_\kappa) \cdot \mathbf{a}, \tag{4.59}$$

so that if $C \neq 0$, we may eliminate Θ from (4.56) and obtain again a classical *propagation condition:*

$$(\hat{\mathbf{A}}(\mathbf{n}_\kappa) - U_\kappa{}^2\mathbf{1})\mathbf{a} = \mathbf{0}, \tag{4.60}$$

where

$$\hat{\mathbf{A}}(\mathbf{n}_\kappa) \equiv \tilde{\mathbf{A}}(\mathbf{n}_\kappa) + \frac{1}{C}\mathbf{L}(\mathbf{n}_\kappa) \otimes \mathbf{L}(\mathbf{n}_\kappa). \tag{4.61}$$

† That the tensor $\tilde{\mathbf{A}}$ is symmetric follows here from the assumption $(4.37)_1$. In all the special theories so far worked out (4.37) is a *proved* consequence of the stress relation $(3.14)_2$; as we have remarked above, it may be so always.

The symmetric tensors $\tilde{\mathbf{A}}(\mathbf{n}_\kappa)$ and $\hat{\mathbf{A}}(\mathbf{n}_\kappa)$ are called the *homothermal and homocaloric acoustic tensors*†, respectively. The direction of a possible kinematic amplitude $\mathbf{a}$ of an acceleration wave is called an *acoustic axis;* the square of a corresponding speed of propagation is called an *acoustic number.* Since $\tilde{\mathbf{A}}$ and $\hat{\mathbf{A}}$ are symmetric, each has at least one real orthogonal triad of acoustic axes, and the acoustic numbers are all real. Thus the results so far obtained may be expressed as follows:

Generalized Fresnel-Hadamard-Duhem Theorem‡: *The free-energetic functional* $\mathbb{F}$ *determines for a given history* (λ^t, γ^t) *unique homothermal and homocaloric acoustic tensors* $\tilde{\mathbf{A}}(\mathbf{n}_\kappa)$, $\hat{\mathbf{A}}(\mathbf{n}_\kappa)$. *Either of these tensors determines the other through* (4.61). *The proper directions of these tensors are the homothermal and homocaloric acoustic axes; of each of these, there exists at least one real orthogonal triad. The corresponding proper numbers are the homothermal and homocaloric acoustic numbers; these are always real.*

The authors of recent researches on waves in bodies whose particles are of somewhat more special kinds, namely, elastic materials and materials with fading memory or quasi-elastic response, have proved or indicated corresponding generalizations of the *equivalence theorems* of acoustics: (I) The propagation condition for all weak waves is the same as that for acceleration waves, *i.e.*, (4.58) for homothermal waves and (4.60) for homocaloric ones. (II) The propagation condition for plane infinitesimal sinusoidal free oscillations according to linearized differential equations is the same as that for acceleration waves. I do not think it would be difficult to extend these theorems to the present more general theory. Thus the

† In my earlier work and in much of the literature this term is applied to the proportional tensors $\tilde{\mathbf{Q}}(\mathbf{n})$ and $\tilde{\mathbf{Q}}(\mathbf{n})$, defined as follows:

$$\tilde{\mathbf{Q}}(\mathbf{n}) = \rho \frac{U^2}{U_\kappa{}^2} \mathbf{A}(\mathbf{n}_\kappa), \qquad \hat{\mathbf{Q}}(\mathbf{n}) = \rho \frac{U^2}{U_\kappa{}^2} \hat{\mathbf{A}}(\mathbf{n}_\kappa),$$

in terms of which the propagation conditions are

$$(\tilde{\mathbf{Q}}(\mathbf{n}) - \rho U^2 \mathbf{1})\mathbf{a} = \mathbf{0}, \qquad (\hat{\mathbf{Q}}(\mathbf{n}) - \rho U^2 \mathbf{1})\mathbf{a} = \mathbf{0}.$$

‡ This theorem, one of the most sweeping in all of mechanics, is distilled by generalization from the work of many authors:

- NEWTON, EULER, LAGRANGE, LAPLACE (Small oscillations in perfect fluids)
- FRESNEL, POISSON (Small oscillations in linear elasticity)
- HUGONIOT, CHRISTOFFEL, HADAMARD (Singular surfaces in linear and non-linear fluid mechanics and elasticity)
- DUHEM (Thermodynamic effects)
- TRUESDELL (Compact and inclusive proof, special features for isotropic materials)
- VARLEY (Materials of integral type)
- COLEMAN & GURTIN (Simple materials, thermodynamic effects)
- WANG & BOWEN (Quasi-elastic response), and others.

behavior of acceleration waves, to the extent that it is expressed by the propagation conditions (4.58) and (4.60), is typical for many kinds of weak disturbances and deserves to be regarded as expressing the *theory of sound*.

We may summarize the results so far in other terms: The laws of propagation of homothermal or homocaloric acceleration waves in a body whose particles obey Axioms 1, 2^+, 3^+, 7, and 8 are, at any particle and place and time, *the same as those in a certain hyperelastic body*, but what that body is depends not only upon the free-energetic functional $\mathbb{F}$ but also upon the history of deformation and temperature. The general theory does not tell us how to *solve* specific problems of wave motion in materials with memory, but it does enable us to *describe* accurately some properties of any wave that may travel through such a body.

It should come as no surprise that the nature of possible wave motions depends very strongly upon thermodynamic conditions. Indeed, in ordinary gas dynamics, where the stress relation is of the form

$$\mathbf{T} = -\hat{p}(\rho, \eta)\mathbf{1} = -\tilde{p}(\rho, \theta)\mathbf{1}, \tag{4.62}$$

substitution in (4.52) and (4.61) yields

$$\begin{aligned} \frac{U_\kappa{}^2}{U^2}\hat{\mathbf{A}}(\mathbf{n}) &= (\partial_\eta \hat{p})\mathbf{n} \otimes \mathbf{n}, \\ \frac{U_\kappa{}^2}{U^2}\tilde{\mathbf{A}}(\mathbf{n}) &= (\partial_\theta \tilde{p})\mathbf{n} \otimes \mathbf{n}, \end{aligned} \tag{4.63}$$

so that from (4.58) and (4.60) we may read off the classic result: All homocaloric or homothermal waves are longitudinal, and

$$U^2 = \begin{cases} \partial_\eta \hat{p} & \text{for homocaloric waves,} \\ \partial_\theta \tilde{p} & \text{for homothermal waves.} \end{cases} \tag{4.64}$$

We know not only that these speeds pertain to different physical circumstances, according as the fluid does or does not conduct heat, but also that there is definite thermodynamic relation between them, namely,

$$\frac{U^2_{\text{homocaloric}}}{U^2_{\text{homothermal}}} = \frac{\kappa_p}{\kappa_v} \tag{4.65}$$

where κ_p and κ_v are the specific heats at constant pressure and constant density, respectively. Now in the general theory we have been discussing, $(4.34)_2$ and (4.57) show that θC is in effect the specific heat at constant deformation $\mathbf{F}$. Thus (4.61) contains (4.65) as a special case. Moreover, if we assume, as is reasonable from the interpretation as a specific heat, that $C > 0$, then (4.61) shows that $\hat{\mathbf{A}}(\mathbf{n}_\kappa)$ is positive-semi-definite if $\tilde{\mathbf{A}}(\mathbf{n}_\kappa)$ is. Hence *if all the homothermal wave speeds are real, so are all the homocaloric ones.*

Further, suppose that $\mathbf{a}$ is a possible homothermal amplitude corresponding to the wave normal $\mathbf{n}$, and let U^2 be the corresponding squared speed. Then by (4.61)

$$\hat{\mathbf{A}}\mathbf{a} = U_\kappa{}^2\mathbf{a} + \frac{1}{C}(\mathbf{L}\cdot\mathbf{a})\mathbf{L}. \tag{4.66}$$

Therefore, a homothermal axis is a homocaloric axis if and only if it is perpendicular or parallel to $\mathbf{L}$. For an Eulerian gas, $\mathbf{L}$ is parallel to $\mathbf{n}$, and since $\mathbf{a}$ must also be parallel to $\mathbf{n}$, this condition is satisfied. Both kinds of waves, in fact, are longitudinal. In general, however, the two sets of acoustic axes are different.

We return now to the condition $(4.36)_3$, which has not been used so far. For a homothermal wave, substitution of $(4.36)_2$ and $(4.32)_2$ into $(4.36)_3$ yields a unique value for $D_\gamma\mathbb{I}[\dot\gamma]\cdot\mathbf{n}_\kappa$ in terms of $\mathbf{n}_\kappa$, U_κ, and $\mathbf{a}$. This may be related to the amplitude defined by [Grad γ]. Thus, in general, a homothermal wave will carry certain discontinuities in the second gradient of the temperature.

For waves that are not homothermal, *e.g.*, for homocaloric waves, $(4.31)_2$ does not hold, and neither does $(4.32)_2$, so we cannot easily proceed further. Some information can be extracted from Fourier's condition (4.12), which, as may be seen from the example furnished by (4.20), relates the wave normal $\mathbf{n}$ to the jump in γ. From Fourier's condition may be derived also a second-order condition of compatibility†, relating the rate of growth of $[\gamma]$ to the curvature of the wave, but it is complicated in form and seems to be inconclusive. While for homocaloric waves the propagation condition (4.60) has been derived, for more general waves that are not homothermal no propagation condition of classical form can hold, and they are definitely not to be described in terms of the properties of an associated hyperelastic body but require a much more delicate analysis, into which we shall not enter here. In fact, homothermal and homocaloric waves may be the most important ones in practice, as we shall now see.

In gas dynamics, we have all been shown a classic argument to persuade us that the true speed of sound is the "Laplacian" or homocaloric one. This argument rests on assuming that the heating flux $\mathbf{h}$ vanishes. If, on the other hand, the gas conducts heat according to Fourier's law, with positive conductivity, the speed of sound is the "Newtonian" or homo-

† *Cf.* Sections 176 and 181 of *The Classical Field Theories*, in FLÜGGE's *Handbuch der Physik*, vol. III/1, Springer-Verlag, Berlin-Göttingen-Heidelberg, 1960.

thermal one. These arguments are easily generalized to materials with memory.

In fact, the first one is trivial, since if $\mathbf{h} \equiv \mathbf{0}$, $(4.32)_2$ shows that $[\dot{\eta}] = 0$, so the results for homocaloric waves follow. Thus we have established the following theorem: *In a non-conducting body, every acceleration wave is homocaloric*, so that if heat conduction is negligible, the appropriate acoustic tensor is $\hat{\mathbf{A}}$, the homocaloric one.

To find the effect of heat conduction, let us reconsider for a moment the classical Fourier-Duhamel theory, defined by (4.20). In that theory the conductivity $\mathbf{C}$ is usually required to be positive-definite (*cf.* Lectures 2 and 7). If so, it is invertible, and (4.20) yields

$$\operatorname{grad} \theta = \mathbf{C}^{-1}\mathbf{h}. \tag{4.67}$$

Consequently, if $[\mathbf{h}] = \mathbf{0}$ at an acceleration wave, then $[\operatorname{grad} \theta] = \mathbf{0}$, and hence the wave is homothermal†. This same result can be extended to the more general materials considered here. Indeed, we need only assume that the functional $\mathbb{I}$ remain defined‡ for small local jump continuations of $\boldsymbol{\gamma}$ (*cf.* (4.40)). Thus we may regard $\mathbb{I}$ as a functional of $\boldsymbol{\lambda}^t$ and $\boldsymbol{\gamma}_+{}^t$, where $\boldsymbol{\gamma}_+{}^t$ is the past history of $\boldsymbol{\gamma}$:

$$\boldsymbol{\gamma}_+{}^t(s) \equiv \boldsymbol{\gamma}^t(s), \qquad 0 < s < \infty, \tag{4.68}$$

provided the present value $\boldsymbol{\gamma}(t)$ be admitted as a parameter:

$$\mathbb{I}(\boldsymbol{\lambda}^t, \boldsymbol{\gamma}^t) = \mathbb{I}(\boldsymbol{\lambda}^t, \boldsymbol{\gamma}_+{}^t; \boldsymbol{\gamma}(t)). \tag{4.69}$$

† Really, it is not necessary to require that $[\mathbf{h}] = \mathbf{0}$, since Fourier's condition (4.12) suffices. More generally, suppose $\mathbf{h} = \mathbf{h}(\boldsymbol{\gamma})$, where the function $\mathbf{h}(\cdot)$ is not necessarily a linear one, and where the arguments θ and $\mathbf{F}$ are not written. Fourier's condition (4.12) becomes

$$\mathbf{n} \cdot \{\mathbf{h}(\boldsymbol{\gamma} + [\boldsymbol{\gamma}]) - \mathbf{h}(\boldsymbol{\gamma})\} = 0.$$

By $(4.10)_1$, $[\boldsymbol{\gamma}]$ is normal to $\mathcal{S}$, say $[\boldsymbol{\gamma}] = G\mathbf{n}$, so this condition becomes

$$(*) \qquad G\mathbf{n} \cdot \{\mathbf{h}(\boldsymbol{\gamma} + G\mathbf{n}) - \mathbf{h}(\boldsymbol{\gamma})\} = 0.$$

If $G = 0$, this condition is satisfied trivially, while if $G \neq 0$, it may be inconsistent with the nature of the material. A *definite conductor* is a material such that

$$\mathbf{m} \cdot \{\mathbf{h}(\boldsymbol{\gamma} + \mathbf{m}) - \mathbf{h}(\boldsymbol{\gamma})\} \neq 0$$

unless $\mathbf{m} = \mathbf{0}$. (In Fourier's theory, this requirement is weaker than the usual one, but stronger than demanding that $\mathbf{C}$ be non-singular.) For a definite conductor, the condition (*) is violated unless $G = 0$. Thus, again, every acceleration wave in a strong conductor is homothermal.

This definition, argument, and result extend immediately to materials with quasi-elastic response.

‡ We need not require that such a continuation be a process, but merely that the one functional $\mathbb{I}$ remain defined and smooth under it.

A material is then a *definite conductor* if for fixed λ^t and $\gamma_+{}^t$ the function $\mathbb{I}(\lambda^t, \gamma_+{}^t; \gamma(t))$ is uniquely invertible for $\gamma(t)$. The above argument goes through unchanged, and we conclude that *in a definite conductor, every acceleration wave is homothermal.*

The two special cases just set up for study are mutually exclusive, of course: A non-conductor is never a definite conductor, and conversely.

The analysis so far is in the nature of extending classical lines of reasoning to circumstances far more general than the old theories ever allowed. However, inclusion and unification of diverse known results are not enough when a new theory is proposed; it must obtain new results as well, predict new effects that lie outside the range of the classical ideas it unifies. In fact, the theory of waves in materials with memory has already predicted such an effect, namely, the existence of a critical amplitude of plane waves. In classical linear acoustics, a plane wave travels unchanged in amplitude forever, and all its parts travel at the same speed. In acoustical experience, weak sounds die out in time, but strong sounds may reinforce so as to become loud noises. Classical studies of acoustics have treated these two phenomena separately. First, they have considered linearized theories of dissipation (obtained from the Navier-Stokes and Boltzmann theories) in which all sounds die out as they progress, and all kinds and parts of waves travel at a common speed. Second, they have considered wave motions according to the non-linear gas dynamics of EULER and HADAMARD, in which there is no dissipation, but reinforcement and shock waves can and do occur. Results of this latter kind have been obtained also, very recently, for the more elaborate types of waves possible according to the theory of finite strain of elastic solids. In a unified theory, including both dissipation and non-linear response, both effects must occur, and there must be a *critical amplitude* a_0 such that if $|\mathbf{a}| < a_0$, the wave dies out because the effects of dissipation predominate, while if $|\mathbf{a}| > a_0$, the wave explodes after a finite time because the effects of reinforcement predominate. In a monumental analysis of plane waves in the theory of materials with fading memory in COLEMAN & NOLL's sense, COLEMAN & GURTIN in 1965 calculated this critical amplitude†. They

† B. D. COLEMAN & M. E. GURTIN, "Waves in materials with memory. III. Thermodynamic influences on the growth and decay of acceleration waves," *Arch. Rational Mech. Anal.*, **19**:266–298 (1965). Reprinted in *Wave Propagation in Dissipative Materials*, Springer-Verlag, New York 1965. Correction, *Arch. Rational Mech. Anal.*, **29**:401 (1968).

Special cases, and more particularly limit cases, of (4.70) had appeared in the earlier literature, which may be summarized roughly as follows:

Gas dynamics: classical authors and T. Y. THOMAS (1957)
Linear elasticity: T. Y. THOMAS (1957)
Hypo-elasticity: NARIBOLI (1964), VARLEY & DUNWOODY (1965)

found the following expressions for the amplitude $a(t)$ of a plane acceleration wave propagating into a definite conductor of infinite extent and uniform constant deformation gradient and temperature:

$$a(t) = \frac{\lambda}{[(\lambda/a(0)) - 1]e^{\mu t} + 1}, \tag{4.70}$$

where λ and μ are constants depending on the material and the past history of deformation and temperature at the particle occupying the place where the wave now is. In a definite conductor, plausible assump-

---

Finite elasticity: W. A. GREEN (1964, 1965)
Materials of the integral type: VARLEY (1965), DUNWOODY & DUNWOODY (1965)

Except for the case of classical gas dynamics, thermodynamic influences were not considered.

Writers on gas dynamics and acoustic relaxation, when they come to consider any effect not covered by the Navier-Stokes equations, prefer to abandon the direct methods of continuum mechanics. Neither, usually, do they resort to real statistical mechanics. Rather, motivated by rough sketches of the molecular state of affairs, they lump the effects of memory in a few scalar parameters, which they call "internal state variables". Shortly after COLEMAN & GURTIN had published their general formulae (4.70) and (4.71), results of the same forms but with special values of λ and μ were obtained in the "molecular relaxation" theory by W. BÜRGER, "Entstehung von Verdichtungsstössen in Gasen mit thermodynamischer Relaxation," *Z. Angew. Math. Mech.*, **46**:T187–T189 (1966), and more generally, by B. D. COLEMAN & M. E. GURTIN "Growth and decay of discontinuities in fluids with internal state variables," *Phys. Fluids*, **10**:1454–1458 (1967). *Cf.* also BÜRGER's Darmstadt dissertation, *Verdichtungswellen endlicher Amplitude in Gasen mit thermodynamischer Relaxation*, 1967, and the paper by B. H. S. RARITY, "On the breakdown of characteristic solutions in flows with vibrational relaxation," *J. Fluid Mech.*, **27**:49–57 (1967).

Theories of this kind are nothing but special cases of the thermodynamic theory of simple materials with fading memory, as has been shown by the explicit calculations of B. D. COLEMAN & M. E. GURTIN, "Some properties of gases with vibrational relaxation when viewed as materials with memory," *Meccanica*, **2**:135–140 (1967), and, less generally, W. BÜRGER, "Ein Gas mit thermodynamischer Relaxation als Beispiel für Materialien mit schwindendem Gedächtnis," *Z. Angew. Math. Mech.*, **47**:T139–T140 (1967). As the titles of these two papers suggest, the results (4.70) and (4.71) apply immediately to the gas with vibrational relaxation, in which the memory functionals, and consequently also the parameters λ and μ, have special forms. While a well known writer on gas dynamics has proclaimed the converse, namely, that the theory of fading memory is only a more complicated formalism for the theory of molecular relaxation, this is so obviously false that mere statement of it is its own refutation. However, since errors seem to propagate nowadays more readily than truths (the medium being less resistant, or, to use an archaic term, more sympathetic, because more like in kind to that which perturbs it), this footnote must close with a reminder that (4.70) holds also in bodies which, unlike the gas with internal state variables, can support shear stress.

tions lead to the inequality $\mu > 0$. By (4.70), the critical amplitude is then $|\lambda|$. That is, if $|a(0)| < |\lambda|$, the amplitude $a(t)$ decays steadily to 0 as $t \to \infty$. If $|a(0)| > |\lambda|$, then $|a(t)| \to \infty$ as $t \to$ the finite time t_∞ given by

$$t_\infty = -\frac{1}{\mu}\log\left(1 - \frac{\lambda}{a(0)}\right). \tag{4.71}$$

In the original memoir, the meanings of λ and μ in terms of material response are explained, and in a later work the results have been extended to a much more general class of materials called "Maxwellian"†. For the purpose of this lecture, it is enough to make it clear that the theory yields two essentially different types of plane waves, those that are damped out and those that are reinforced, according to the initial amplitude and the prevailing conditions of deformation and caloric, as well as the nature of the material and the history of deformation and temperature it has undergone‡.

I have tried in this lecture to give some idea of recent work on the nature of wave motions in dissipative materials. I must beg pardon if to some of you my text may have seemed like a telegram. To present in any detail the marvelous growth in our knowledge of wave propagation in the last three years, I should have had to use all of the lectures in this series. One thing, I think, is clear. Thermodynamics, after its fitful drowse of nearly a century, has awakened and come of age, joining mechanics and electromagnetism as a great field theory in which non-trivial problems may be set and solved.

In the next two lectures we shall see how the basic notion of irreversibility in continuum mechanics can itself be extended so as to apply to mixtures.

† B. D. Coleman, J. Greenberg, & M. E. Gurtin, "Waves in materials with memory. V. On the amplitude of acceleration waves and mild discontinuities," *Arch. Rational Mech. Anal.*, **12**:333–354 (1966).

‡ Similar but of course more complicated results hold for cylindrical and spherical waves. Beyond the results known for some time in classical gas dynamics and in linear elasticity, there is an important paper on this problem in the purely mechanical theory of materials of the integral type by E. Varley, "Acceleration fronts in visco-elastic materials," *Arch. Rational Mech. Anal.*, **19**:215–225 (1965). A more detailed and specific treatment for curved waves in a homogeneous isotropic body at rest subject to hydrostatic pressure has been constructed by P. J. Chen, "The growth of acceleration waves of arbitrary form in homogeneously deformed elastic materials," *Arch. Rational Mech. Anal.*, **30**:81–89 (1968). Doubtless his result could be extended so as to take account of possible changes of temperature.

Note added in March, 1969. The extension has been made by Chen in his later paper, "Thermodynamic influences on the propagation and the growth of acceleration waves in elastic materials," *Arch. Rational Mech. Anal.*, **31**:228–254 (1968), **32**:400–401 (1969).

LECTURE 5

THERMODYNAMICS OF DIFFUSION

So as to describe the interactions of bodies composed of identifiable constituents, we set up the mathematical theory of *mixtures*. The theory aims to represent the phenomena of diffusion, dissociation, combination, and chemical reaction in the broadest sense. It finds long antecedence in the special theories of diffusion used by chemists and physicists and in the theories of homogeneous chemical reactions, not to mention special static and kinetic theories of heterogeneous media, but a general framework for all such theories seems first to have been laid down in a note of mine† in 1957. It has been extended by KELLY‡ so as to include electromagnetic effects, couple stresses, and singular surfaces, but today I shall need only its simplest mechanical and thermodynamic aspects.

We consider a sequence of bodies $\underset{\mathfrak{a}}{\mathcal{B}}$, $\mathfrak{a} = \mathit{1}, \mathit{2}, \ldots, \mathfrak{f}$, all of which we suppose able to occupy regions of space, and indeed simultaneously. Thus if $\underset{\mathfrak{a}}{\mathbf{X}}$ is the place taken by a particle of $\underset{\mathfrak{a}}{\mathcal{B}}$ in some reference configuration, the *motion* of $\underset{\mathfrak{a}}{\mathcal{B}}$ is the smooth mapping

$$\mathbf{x} = \underset{\mathfrak{a}}{\mathbf{x}}(\underset{\mathfrak{a}}{\mathbf{X}}, t), \qquad -\infty < t < \infty, \tag{5.1}$$

of $\underset{\mathfrak{a}}{\mathcal{B}}$ onto a time-sequence of configurations in space. Each such motion has its own kinematics. We shall use a backward prime to denote the time derivative when $\underset{\mathfrak{a}}{\mathbf{X}}$ is held constant. Thus $\underset{\mathfrak{a}}{\grave{\mathbf{x}}}$ is the *velocity of constituent* $\mathfrak{a}$:

$$\underset{\mathfrak{a}}{\grave{\mathbf{x}}} \equiv \partial_t \underset{\mathfrak{a}}{\mathbf{x}}(\underset{\mathfrak{a}}{\mathbf{X}}, t) = \underset{\mathfrak{a}}{\grave{\mathbf{x}}}(\mathbf{x}, t); \tag{5.2}$$

likewise its *acceleration* is $\underset{\mathfrak{a}}{\grave{\grave{\mathbf{x}}}}$:

$$\underset{\mathfrak{a}}{\grave{\grave{\mathbf{x}}}} \equiv \partial_t{}^2 \underset{\mathfrak{a}}{\mathbf{x}}(\underset{\mathfrak{a}}{\mathbf{X}}, t) = \underset{\mathfrak{a}}{\grave{\grave{\mathbf{x}}}}(\mathbf{x}, t). \tag{5.3}$$

Each body $\underset{\mathfrak{a}}{\mathcal{B}}$ has its own mass and consequently its *mass density* $\underset{\mathfrak{a}}{\rho}$ in the configuration $\underset{\mathfrak{a}}{\mathbf{x}}$ at time t.

We shall consider only places $\mathbf{x}$ in space that are taken simultaneously

† My work is summarized in Sections 158, 159, 215, 254, 255, 259, 261, and 295 of *The Classical Field Theories*, in FLÜGGE's *Handbuch der Physik*, vol. III/1, Springer-Verlag, 1960.

‡ P. D. KELLY, "A reacting continuum," *Intern. J. Eng. Sci.*, **2**:129–153 (1964).

by one particle from each of the $\mathfrak{f}$ constituent bodies. The total *mass density* ρ is defined as the sum of the peculiar densities:

$$\rho \equiv \Sigma \underset{\mathfrak{a}}{\rho}. \tag{5.4}$$

Here and henceforth, Σ stands for summation from $\mathfrak{a} = 1$ to $\mathfrak{a} = \mathfrak{f}$. The *concentration* or *mass fraction* $\underset{\mathfrak{a}}{c}$ of the constituent $\mathfrak{a}$ is the dimensionless ratio given by

$$\underset{\mathfrak{a}}{c} \equiv \frac{\underset{\mathfrak{a}}{\rho}}{\rho}. \tag{5.5}$$

Thus

$$\Sigma \underset{\mathfrak{a}}{c} = 1. \tag{5.6}$$

The mixture may be regarded as being in motion also as a single body. Quantities associated with it are of two kinds. First, there are densities of additive set functions. These are simply the sums, weighted by the concentrations, of the densities of the constituents. For example, for the densities of linear momentum and rotational momentum we lay down the following definitions:

$$\begin{aligned} \dot{\mathbf{x}} &\equiv \Sigma \underset{\mathfrak{a}}{c}\underset{\mathfrak{a}}{\dot{\mathbf{x}}}, \\ \mathbf{p} \wedge \dot{\mathbf{x}} &\equiv \Sigma \mathbf{p} \wedge \underset{\mathfrak{a}}{c}\underset{\mathfrak{a}}{\dot{\mathbf{x}}}, \end{aligned} \tag{5.7}$$

where $\mathbf{p}$ is the position vector of $\mathbf{x}$ with respect to some origin. The first of these states that the total linear momentum of the mixture in any region is the sum of the linear momenta of the constituents presently occupying that region. The second, which in the present simple theory is a consequence of the first, asserts a like conclusion for the rotational momentum. The first gives, incidentally, a definition of the motion of the mixture if we choose to interpret $\dot{\mathbf{x}}$ as a velocity. KELLY has proposed a more general form for the second, including spin momentum, but we shall not need it here.

Sometimes it lightens the wording if we call *peculiar*† those quantities associated with the individual constituents $\underset{\mathfrak{a}}{\mathfrak{B}}$ and hence denoted by a small German letter written below them, *e.g.*, $\underset{\mathfrak{a}}{\rho}$, $\underset{\mathfrak{a}}{c}$. Thus $\underset{\mathfrak{a}}{\dot{\mathbf{x}}}$ is the *peculiar velocity* of constituent $\mathfrak{a}$, *etc.*

† There seems to be no appropriate term here that does not have some connotation unwanted for the present subject. The term "partial", used by some authors, suggests to those trained in physical chemistry some partial derivative of some potential function. The term "peculiar" seems to give the meaning best, but in the kinetic theory of gases it is used in the sense of "relative to the mean".

Each constituent suffers actions of two kinds: the prescribed actions at a distance, represented here by the densities of *body force* $\underset{a}{\mathbf{b}}$ and *body heating* $\underset{a}{s}$, and the contact actions, represented here by the *stress* $\underset{a}{\mathbf{T}}$ and *heating flux* $\underset{a}{\mathbf{h}}$. The *inner parts* of the resultant actions on the mixture are the sums of these†—concentration-weighted sums for the former, since they are densities with respect to mass, and simple sums for the latter, since they are densities with respect to area:

$$\begin{aligned} \mathbf{b}_{\mathrm{I}} &\equiv \Sigma \underset{a}{c}\underset{a}{\mathbf{b}}, \qquad & s_{\mathrm{I}} &\equiv \Sigma \underset{a}{c}\underset{a}{s}, \\ \mathbf{T}_{\mathrm{I}} &\equiv \Sigma \underset{a}{\mathbf{T}}, \qquad & \mathbf{h}_{\mathrm{I}} &\equiv \Sigma \underset{a}{\mathbf{h}}. \end{aligned} \tag{5.8}$$

Likewise, the inner energetic and caloric of the mixture are defined in such a way as to render the internal energy and calory of the mixture in a region the sums of the internal energies and calories, respectively, of the parts of the constituent bodies $\underset{a}{\mathcal{B}}$ occupying that same region, *viz*

$$\varepsilon_{\mathrm{I}} \equiv \Sigma \underset{a}{c}\underset{a}{\varepsilon}, \qquad \eta \equiv \Sigma \underset{a}{c}\underset{a}{\eta}. \tag{5.9}$$

These inner parts do not account for all the energy and all the body and contact actions on the mixture, not only since the relative motions of the constituents may themselves carry momentum and energy about, but also since mass, momentum, and energy may be given to one constituent or taken from it by the chemical reactions and physical transfers occurring.

That we may go on from here with open eyes, we lay down three metaphysical principles:

1. *All properties of the mixture must be mathematical consequences of properties of the constituents.*
2. *So as to describe the motion of a constituent, we may in imagination isolate it from the rest of the mixture, provided we allow properly for the actions of the other constituents upon it.*
3. *The motion of the mixture is governed by the same equations as is a single body.*

The first principle asserts, roughly, that the whole is no more than the sum of its parts, and the third, that in its motion as a whole a body does not know whether it is a mixture or not. The second is an extension of the familiar principle of solidification, by which the parts of a body

† GURTIN informs me that he, NOLL, & W. WILLIAMS have set up an axiomatic structure for the mechanics and thermodynamics of mixtures in which results of this type emerge naturally.

occupying different regions of space are imagined cut asunder in geometry but united in physics by suitable forces or energies. Here we distinguish in imagination different constituents occupying the same region of space.†

These statements acknowledge our procedure. Many scientists are taught not to make any metaphysical confession, or at least not to recognize such as they do make. Scientists of that kind may erase from their magnetic tapes my last few sentences and then follow my procedure without specifying it.

For a region $\mathcal{V}$ of space, we may consider the actions on the part of the constituent body $\underset{\mathfrak{a}}{\mathcal{B}}$ presently occupying $\mathcal{V}$ and calculate the rates of growth of mass, linear momentum, rotational momentum, and total energy within it. We assume, in the spirit of continuum mechanics, that these rates of growth are absolutely continuous functions of the mass of the mixture, and we name their densities as follows:

$\underset{\mathfrak{a}}{\overset{+}{c}}$ growth of mass,

$\underset{\mathfrak{a}}{\overset{+}{\mathbf{m}}}$ growth of linear momentum,

$\underset{\mathfrak{a}}{\overset{+}{\mathbf{M}}}$ growth of rotational momentum,

$\underset{\mathfrak{a}}{\overset{+}{e}}$ growth of energy.

† Motivation, if it is needed, may be found both in the classical mechanics of mass-points and in the kinetic theory of gases. In the former, the concept of force exerted by A on B is primitive and not subject to redefinition. In the kinetic theory, the quantities $\underset{\mathfrak{a}}{\mathbf{b}}$, $\underset{\mathfrak{a}}{\mathbf{T}}$, $\underset{\mathfrak{a}}{s}$, and $\underset{\mathfrak{a}}{\mathbf{h}}$ are *uniquely defined* in terms of the forces exerted on the molecules of constituent $\mathfrak{a}$ and the momentum and energy carried by those molecules.

Obviously, we should be able to divide any single body arbitrarily into two constituents, *e.g.*, by regarding the particle $\mathbf{X}$ with mass density $\rho(\mathbf{X}, t)$ as being in fact two particles $\underset{1}{\mathbf{X}}$ and $\underset{2}{\mathbf{X}}$ with present densities $\underset{1}{\rho}(\mathbf{X}, t)$ and $\underset{2}{\rho}(\mathbf{X}, t)$ such that $\underset{1}{\rho} + \underset{2}{\rho} = \rho$, and be unable to distinguish any change in the resulting motion. So as to stay forever united, these two purely ideal constituents must exert suitable forces on each other. When we come to extend this idea to mixtures of identifiable constituents, we recognize a like arbitrariness in the division, even in terms of ordinary chemistry. For example, is a mixture of water and alcohol in a whisky conduit to be regarded as a mixture of those two constituents, or of three ideal substances labeled H, O, and C that are bound together inseparably in two particular ways? Clearly it makes no difference.

I proposed the above principles over ten years ago. The second one, then phrased in inadvertently special terms, is here recast.

When we come to the theory of constitutive equations, the principle of equipresence frees us from having to conjecture in advance differences in response functions arising from different separations of a mixture into parts. According to that principle, such differences should *follow* from the physical and chemical properties laid down for the mixture, not having to be left to "intuition".

Accordingly, the four equations of balance for the constituent $\mathfrak{a}$ are:

$$
\begin{aligned}
\int \underset{\mathfrak{a}}{\rho} \overset{+}{c}\, dv &\equiv \Big(\int \underset{\mathfrak{a}}{\rho}\, dv\Big)^{\backprime} \\
\int \rho \underset{\mathfrak{a}}{\overset{+}{\mathbf{m}}}\, dv &\equiv \Big(\int \underset{\mathfrak{a}}{\rho} \underset{\mathfrak{a}}{\dot{\mathbf{x}}}\, dv\Big)^{\backprime} - \oint \underset{\mathfrak{a}}{\mathbf{T}}\mathbf{n}\, dA - \int \underset{\mathfrak{a}}{\rho} \underset{\mathfrak{a}}{\mathbf{b}}\, dv, \\
\int \rho\big(\underset{\mathfrak{a}}{\overset{+}{\mathbf{M}}} + \mathbf{p} \wedge \underset{\mathfrak{a}}{\overset{+}{\mathbf{m}}}\big)\, dv &\equiv \Big(\int \mathbf{p} \wedge (\underset{\mathfrak{a}}{\rho} \underset{\mathfrak{a}}{\dot{\mathbf{x}}})\, dv\Big)^{\backprime} - \oint \mathbf{p} \wedge \mathbf{T}\mathbf{n}\, dA - \int \mathbf{p} \wedge \underset{\mathfrak{a}}{\rho} \underset{\mathfrak{a}}{\mathbf{b}}\, dv, \\
\int \rho \underset{\mathfrak{a}}{\overset{+}{e}}\, dv &\equiv \Big(\int \underset{\mathfrak{a}}{\rho}(\underset{\mathfrak{a}}{\varepsilon} + \tfrac{1}{2}\underset{\mathfrak{a}}{\dot{x}}^2)\, dv\Big)^{\backprime} - \oint \underset{\mathfrak{a}}{\dot{\mathbf{x}}} \cdot \underset{\mathfrak{a}}{\mathbf{T}}\mathbf{n}\, dA - \oint \underset{\mathfrak{a}}{\mathbf{h}} \cdot \mathbf{n}\, dA \\
&\qquad - \int \underset{\mathfrak{a}}{\rho} \underset{\mathfrak{a}}{\dot{\mathbf{x}}} \cdot \underset{\mathfrak{a}}{\mathbf{b}}\, dv - \int \underset{\mathfrak{a}}{\rho} \underset{\mathfrak{a}}{s}\, dv.
\end{aligned}
\tag{5.10}
$$

In these equations, $\int$ denotes integration over the volume $\mathcal{V}$, while $\oint$ denotes integration over its boundary $\partial\mathcal{V}$, and a backward prime denotes the time derivative when the set of particles of $\underset{\mathfrak{a}}{\mathcal{B}}$ occupying $\mathcal{V}$ is held constant.

The theory of mixtures is more complicated than the theory of a single body but not different in kind. In elementary mechanics, when we speak of a system of forces exerted by a set of bodies on each other, we mean *the* forces, and we use them as such, writing equations in which they occur with values assumed unique, without comment. In statics, for example, *the* forces in a system of forces are primitive elements, to which we assign not only some physical sense but also individuality in the mathematical theory. Likewise, in continuum mechanics we speak of *the* stress tensor and *the* heating flux, since otherwise it would be senseless to write equations delivering unique values for them. In just the same way, the theory of mixtures introduces *the* peculiar stress $\underset{\mathfrak{a}}{\mathbf{T}}$, *the* peculiar heating flux $\underset{\mathfrak{a}}{\mathbf{h}}$, *etc.*, as unique primitive elements in its structure.

It is an easy matter, if we assume the fields occurring be sufficiently smooth, to apply the divergence theorem and obtain explicit formulae for the four growths:

$$
\begin{aligned}
\rho \underset{\mathfrak{a}}{\overset{+}{c}} &= \underset{\mathfrak{a}}{\dot{\rho}} + \underset{\mathfrak{a}}{\rho} \operatorname{div} \underset{\mathfrak{a}}{\dot{\mathbf{x}}} = \partial_t \underset{\mathfrak{a}}{\rho} + \operatorname{div}(\underset{\mathfrak{a}}{\rho} \underset{\mathfrak{a}}{\dot{\mathbf{x}}}), \\
\rho \underset{\mathfrak{a}}{\overset{+}{\mathbf{m}}} &= \rho \underset{\mathfrak{a}}{\overset{+}{c}} \underset{\mathfrak{a}}{\dot{\mathbf{x}}} + \underset{\mathfrak{a}}{\rho} \underset{\mathfrak{a}}{\ddot{\mathbf{x}}} - \operatorname{div} \underset{\mathfrak{a}}{\mathbf{T}} - \underset{\mathfrak{a}}{\rho} \underset{\mathfrak{a}}{\mathbf{b}}, \\
\rho \underset{\mathfrak{a}}{\overset{+}{\mathbf{M}}} &= \underset{\mathfrak{a}}{\mathbf{T}} - \underset{\mathfrak{a}}{\mathbf{T}}^{\mathsf{T}}, \\
\rho \underset{\mathfrak{a}}{\overset{+}{e}} &= \rho \underset{\mathfrak{a}}{\overset{+}{\mathbf{m}}} \cdot \underset{\mathfrak{a}}{\dot{\mathbf{x}}} + \rho \underset{\mathfrak{a}}{\overset{+}{c}}(\underset{\mathfrak{a}}{\varepsilon} - \tfrac{1}{2}\underset{\mathfrak{a}}{\dot{x}}^2) + \underset{\mathfrak{a}}{\rho} \underset{\mathfrak{a}}{\dot{\varepsilon}} - \operatorname{tr}(\underset{\mathfrak{a}}{\mathbf{T}}^{\mathsf{T}} \operatorname{grad} \underset{\mathfrak{a}}{\dot{\mathbf{x}}}) - \operatorname{div} \underset{\mathfrak{a}}{\mathbf{h}} - \underset{\mathfrak{a}}{\rho} \underset{\mathfrak{a}}{s}.
\end{aligned}
\tag{5.11}
$$

If the mixture reduces to a single constituent, that is, if $\mathfrak{f} = 1$, the left-hand sides of all four equations are zero, and we recover from (5.10) the equations of balance in ordinary continuum mechanics, and from

(5.11), their local equivalents, namely, (2.8), (2.10), and (2.18). In a mixture, we grant that chemical reactions and physical transfers may greaten or lessen the mass, linear momentum, rotational momentum, and energy of any constituent, but we require that these quantities be conserved for the whole mixture. We assume, that is, that the reactions and transfers are exchanges rather than true processes of creation or destruction. By (5.10), this condition is expressed by the following four ***axioms of balance for mixtures:***

$$\sum_{\mathfrak{a}} \overset{+}{c}_{\mathfrak{a}} = 0, \qquad \sum_{\mathfrak{a}} \overset{+}{\mathbf{m}}_{\mathfrak{a}} = \mathbf{0}, \qquad \sum_{\mathfrak{a}} \overset{+}{\mathbf{M}}_{\mathfrak{a}} = \mathbf{0}, \qquad \sum_{\mathfrak{a}} \overset{+}{e}_{\mathfrak{a}} = 0. \tag{5.12}$$

We are now ready to impose the requirement that the mixture as a whole not know whether it be simple or heterogeneous. To do so, we must define appropriately the total stress $\mathbf{T}$, total body force $\mathbf{b}$, total heating flux $\mathbf{h}$, total energetic ε, and total heating supply s. The *diffusion velocity* $\mathbf{u}_{\mathfrak{a}}$ of the constituent $\mathfrak{a}$ is its velocity relative to the mean motion:

$$\mathbf{u}_{\mathfrak{a}} \equiv \dot{\mathbf{x}}_{\mathfrak{a}} - \dot{\mathbf{x}}. \tag{5.13}$$

Since the mixture is conceived as a body in motion with the velocity $\dot{\mathbf{x}}$, the diffusive transport at the velocity $\mathbf{u}_{\mathfrak{a}}$ of the quantities associated with the constituent $\mathfrak{a}$ must be taken into account in calculating the actions on the mixture. The appropriate definitions are easy to motivate and interpret, particularly for anyone familiar with the conceptual basis of the kinetic theory of gas mixtures:

$$\begin{aligned}
\mathbf{T} &\equiv \mathbf{T}_{\mathrm{I}} - \Sigma \rho_{\mathfrak{a}} \mathbf{u}_{\mathfrak{a}} \otimes \mathbf{u}_{\mathfrak{a}}, \\
\mathbf{b} &\equiv \mathbf{b}_{\mathrm{I}}, \\
\varepsilon &\equiv \varepsilon_{\mathrm{I}} + \tfrac{1}{2}\Sigma c_{\mathfrak{a}} u_{\mathfrak{a}}^2, \\
\mathbf{h} &\equiv \mathbf{h}_{\mathrm{I}} + \Sigma[\mathbf{T}_{\mathfrak{a}}^{\mathrm{T}}\mathbf{u}_{\mathfrak{a}} - \rho_{\mathfrak{a}}(\varepsilon_{\mathfrak{a}} + \tfrac{1}{2}u_{\mathfrak{a}}^2)\mathbf{u}_{\mathfrak{a}}], \\
s &\equiv s_{\mathrm{I}} + \Sigma c_{\mathfrak{a}} \mathbf{b}_{\mathfrak{a}} \cdot \mathbf{u}_{\mathfrak{a}}.
\end{aligned} \tag{5.14}$$

With these definitions, the axioms (5.12) become equivalent to the ordinary laws of balance in continuum mechanics, *viz* (in differential form)

$$\dot{\rho} + \rho \operatorname{div} \dot{\mathbf{x}} = 0, \tag{2.3}$$

$$\rho \ddot{\mathbf{x}} = \operatorname{div} \mathbf{T} + \rho \mathbf{b}, \tag{2.8}$$

$$\mathbf{T} = \mathbf{T}^{\mathrm{T}}, \tag{2.10}$$

$$\rho \dot{\varepsilon} = w + \operatorname{div} \mathbf{h} + \rho s, \qquad w = \operatorname{tr}(\mathbf{T}^{\mathrm{T}} \operatorname{grad} \dot{\mathbf{x}}). \qquad (2.18), (2.14)$$

In particular, we notice that while the peculiar stresses $\mathbf{T}_{\mathfrak{a}}$ need not themselves be symmetric, their sum, which is $\mathbf{T}_{\mathrm{I}}$, must be. Moreover, by $(5.14)_5$, the body heating s differs from its inner part s_{I} only if the body forces $\underset{\alpha}{\mathbf{b}}$ are not all alike.

(If we prefer, we could start from (5.12) as an axiom, lay down the above local equations of balance for the mixture as requirements, take only $(5.14)_{2,3,5}$ as definitions, and derive $(5.14)_{1,4}$ as necessary and sufficient conditions, except that $\mathbf{T}$ and $\mathbf{h}$ are determined by this argument only to within an arbitrary solenoidal tensor and vector, respectively.)

The proofs of these results, and of all relations between the properties of the constituents and the properties of the mixture, rest upon a general equation of transport. Suppose that from fields $\underset{\alpha}{\Psi}$ associated with the several constituents we define a field Ψ for the mixture as follows:

$$\Psi \equiv \Sigma \underset{\alpha}{c}\,\underset{\alpha}{\Psi}. \tag{5.15}$$

Then the equation of transport, relating the time derivative $\dot{\Psi}$ of Ψ following the motion of the mixture to the time derivative $\underset{\alpha}{\acute{\Psi}}$ of $\underset{\alpha}{\Psi}$ following the motion of $\underset{\alpha}{\mathcal{B}}$, is the identity

$$\dot{\Psi} = \Sigma \underset{\alpha}{c}\,\underset{\alpha}{\acute{\Psi}} - \frac{1}{\rho}\operatorname{div}\Sigma(\underset{\alpha}{\rho}\,\underset{\alpha}{\Psi}\,\underset{\alpha}{\mathbf{u}}) + \Sigma \underset{\alpha}{\overset{+}{c}}\,\underset{\alpha}{\Psi}, \tag{5.16}$$

as is not difficult to prove.

So far, everything is crystal-clear. Troubles begin, as they always do, when we leave the palace of pure mechanics and break the seal of Sulaiman on the jar of that great ifrit, the entropy or calory.

Until recently, authors in this field were heavily influenced by thermostatics and by the linear "thermodynamics of irreversible processes". They assumed the existence of a local caloric equation of state, whence, with recourse to other special constitutive hypotheses, they calculated the production of caloric. Little came out of the formal structure beyond obvious rearrangements of the assumptions. From the moment COLEMAN & NOLL's fundamental paper on thermo-elasticity appeared, thermodynamics began to expand in all directions, reaching the adulthood of a field theory. For a single medium, I have outlined some of the major results of COLEMAN, GURTIN, and WANG & BOWEN in Lectures 3 and 4. Of course the same approach, based upon requiring the dissipation principle to be satisfied identically, ought to yield definite results for mixtures also. Specifically, a general theory of chemical potentials, combined with restrictions upon the stresses $\underset{\alpha}{\mathbf{T}}$ or at least the inner stress $\mathbf{T}_{\mathrm{I}}$, should follow. So as to obtain it, we need a general framework for the composition of the growth of calory of the mixture from the growths of the calories of the constituents. This framework may be constructed easily† by the methods and concepts I outlined at the beginning for the principles of balance.

Namely, each constituent is assigned its *coldness*‡ $\underset{\alpha}{\vartheta}$ and its *caloric* $\underset{\alpha}{\eta}$. The *growth of caloric* $\overset{+}{\underset{\alpha}{\eta}}$ of constituent α may be defined through an equation of balance parallel to $(5.10)_{1,2,3,4}$, namely:

$$\int \rho \overset{+}{\underset{\alpha}{\eta}}\, dv \equiv \left(\int \underset{\alpha}{\rho}\,\underset{\alpha}{\eta}\, dv\right)^{\cdot} - \oint \underset{\alpha}{\vartheta}\,\underset{\alpha}{\mathbf{h}} \cdot \mathbf{n}\, dA - \int \underset{\alpha}{\vartheta}\,\underset{\alpha}{\rho}\,\underset{\alpha}{s}\, dv, \tag{5.17}$$

where we assume, as in the ordinary theory for a single medium, that the heating flux $\underset{\alpha}{\mathbf{h}}$ and the body heating s determine a flux of caloric $\underset{\alpha}{\vartheta}\underset{\alpha}{\mathbf{h}}$ and a supply of caloric $\underset{\alpha}{\vartheta}\underset{\alpha}{s}$ with a common coldness $\underset{\alpha}{\vartheta}$. The local equivalent of (5.17) is

$$\rho \overset{+}{\underset{\alpha}{\eta}} = \rho \overset{+}{\underset{\alpha}{c}}\,\underset{\alpha}{\eta} + \underset{\alpha}{\rho}\,\dot{\underset{\alpha}{\eta}} - \operatorname{div}\,(\underset{\alpha}{\vartheta}\,\underset{\alpha}{\mathbf{h}}) - \underset{\alpha}{\vartheta}\,\underset{\alpha}{\rho}\,\underset{\alpha}{s}. \tag{5.18}$$

If we sum this equation on α, reduce the result by use of the equation of transport (5.16), and then use the definition (5.9), we obtain

$$\rho\dot{\eta} = \rho \Sigma \overset{+}{\underset{\alpha}{\eta}} + \operatorname{div} \Sigma(\underset{\alpha}{\vartheta}\,\underset{\alpha}{\mathbf{h}} - \underset{\alpha}{\rho}\,\underset{\alpha}{\eta}\,\underset{\alpha}{\mathbf{u}}) + \Sigma \underset{\alpha}{\vartheta}\,\underset{\alpha}{\rho}\,\underset{\alpha}{s}. \tag{5.19}$$

To restrict the several growths of caloric, we now proceed by analogy to the reasoning which led to the conservation axioms (5.12). Regarding the reactions and transfers as exchanges rather than true processes of creation or annihilation, we allow the growth of caloric $\overset{+}{\underset{\alpha}{\eta}}$ of the constituent α to be of any amount so long as the total growth of caloric for the mixture remain non-negative. That is, our ***axiom of dissipation*** is

$$\Sigma \overset{+}{\underset{\alpha}{\eta}} \geqq 0, \tag{5.20}$$

† C. Truesdell, "Sulle basi della termodinamica delle miscele," *Rend. Accad. Naz. Lincei,* (8) **44**:381–383 (1968).

In writing this note I unfortunately forgot that R. Bowen, in work done at the Johns Hopkins University in 1964–1965 and described in a lecture to the Society for Natural Philosophy at Bressanone in June, 1965, had proposed the Clausius-Duhem inequality (5.20). Bowen's work of that time, to the extent it can now be reconstructed, seems not to have reached the simple form (5.21) and its consequences such as (5.29).

‡ So far as I can learn, in my note of 1957 and in an unpublished report by H. Grad written about the same time are the first suggestions that the constituents of a flowing mixture of continua need not have a common temperature. The analysis concerning this fact in my note is restricted to the case when each constituent obeys a caloric equation of state of its own; in Grad's report, to special constitutive equations. Of course, the possibility that the constituents have different temperatures at the same place and time is inherent in Maxwell's kinetic theory of mixtures of monatomic gases, though it seems not to have been developed by writers on that subject.

to be set alongside the four axioms of balance (5.12). From the identity (5.19) it then follows that

$$\rho\dot{\eta} \geqq \operatorname{div} \boldsymbol{\phi} + \rho\sigma, \tag{5.21}$$

where

$$\begin{aligned} \boldsymbol{\phi} &= \Sigma(\underset{\mathfrak{a}}{\vartheta}\underset{\mathfrak{a}}{\mathbf{h}} - \underset{\mathfrak{a}}{\rho}\underset{\mathfrak{a}}{\eta}\underset{\mathfrak{a}}{\mathbf{u}}), \\ \sigma &= \Sigma\underset{\mathfrak{a}}{\vartheta}\underset{\mathfrak{a}}{c}\underset{\mathfrak{a}}{s}. \end{aligned} \tag{5.22}$$

The local inequality (5.21) corresponds to a global Clausius-Duhem inequality of the form

$$\dot{\mathfrak{H}} \geqq \oint \boldsymbol{\phi} \cdot \mathbf{n}\, dA + \int \sigma\, dM. \tag{5.23}$$

In the case of a single medium, (5.22) yields

$$\boldsymbol{\phi} = \vartheta\mathbf{h}, \qquad \sigma = \vartheta s, \tag{5.24}$$

and (5.23) reduces to the classical form (2.50). More generally, with $\mathbf{h}$ and s defined for the mixture by $(5.14)_{4,5}$, the relations (5.24) do not hold.

This fact seems to contradict the third of the guiding principles set at the head of this lecture: "The motion of the mixture is governed by the same equations as is a single body." Indeed, it reflects a certain vagueness in that principle, since "the same equations" are not unequivocally specified. Following the first and second principles strictly has indeed led to a divergence, not so much between the forms of the basic equations and inequality, as in the relation between the mechanisms of growth of energy and of calory. For mixtures, this relation has always been a moot point. We shall come back to it shortly.

Before going further, we record here an expression for $\overset{+}{\underset{\mathfrak{a}}{\eta}}$ reduced by elimination of the body heating $\underset{\mathfrak{a}}{s}$. Namely, if we introduce the free energetic $\underset{\mathfrak{a}}{\psi}$ of the $\mathfrak{a}^{\text{th}}$ constituent:

$$\underset{\mathfrak{a}}{\psi} \equiv \underset{\mathfrak{a}}{\varepsilon} - \underset{\mathfrak{a}}{\theta}\underset{\mathfrak{a}}{\eta}, \tag{5.25}$$

then by substituting $(5.11)_4$ into (5.18) so as to eliminate $\underset{\mathfrak{a}}{s}$ we find that

$$\begin{aligned} \overset{+}{\underset{\mathfrak{a}}{\eta}} = \underset{\mathfrak{a}}{\vartheta}[\overset{+}{\underset{\mathfrak{a}}{e}} - \overset{+}{\underset{\mathfrak{a}}{\mathbf{m}}} \cdot \underset{\mathfrak{a}}{\dot{\mathbf{x}}} - \overset{+}{\underset{\mathfrak{a}}{c}}(\underset{\mathfrak{a}}{\psi} - \tfrac{1}{2}\underset{\mathfrak{a}}{\dot{x}}^2)] + \underset{\mathfrak{a}}{c}\underset{\mathfrak{a}}{\eta}\underset{\mathfrak{a}}{\dot{\vartheta}}/\underset{\mathfrak{a}}{\vartheta} \\ - \underset{\mathfrak{a}}{c}\underset{\mathfrak{a}}{\vartheta}\underset{\mathfrak{a}}{\dot{\psi}} + \frac{1}{\rho}[\underset{\mathfrak{a}}{\vartheta}\, \operatorname{tr}(\underset{\mathfrak{a}}{\mathbf{T}}^{\mathrm{T}} \cdot \operatorname{grad} \underset{\mathfrak{a}}{\dot{\mathbf{x}}}) - \underset{\mathfrak{a}}{\mathbf{h}} \cdot \operatorname{grad} \underset{\mathfrak{a}}{\vartheta}]. \end{aligned} \tag{5.26}$$

The dissipation axiom (5.20) for mixtures requires, then, that

$$\begin{aligned} \Sigma\{\underset{\mathfrak{a}}{\vartheta}[-\overset{+}{\underset{\mathfrak{a}}{e}} + \overset{+}{\underset{\mathfrak{a}}{\mathbf{m}}} \cdot \underset{\mathfrak{a}}{\dot{\mathbf{x}}} + \overset{+}{\underset{\mathfrak{a}}{c}}(\underset{\mathfrak{a}}{\psi} - \tfrac{1}{2}\underset{\mathfrak{a}}{\dot{x}}^2)] + \underset{\mathfrak{a}}{c}(\underset{\mathfrak{a}}{\vartheta}\underset{\mathfrak{a}}{\dot{\psi}} - \underset{\mathfrak{a}}{\eta}\underset{\mathfrak{a}}{\dot{\vartheta}}/\underset{\mathfrak{a}}{\vartheta}) \\ - \frac{1}{\rho}[\underset{\mathfrak{a}}{\vartheta}\, \operatorname{tr}(\underset{\mathfrak{a}}{\mathbf{T}}^{\mathrm{T}} \cdot \operatorname{grad} \underset{\mathfrak{a}}{\dot{\mathbf{x}}}) - \underset{\mathfrak{a}}{\mathbf{h}} \cdot \operatorname{grad} \underset{\mathfrak{a}}{\vartheta}]\} \leqq 0. \end{aligned} \tag{5.27}$$

This, finally, is the ***reduced dissipation inequality for mixtures.*** As in the case of a single medium, the term "reduced" refers to the fact that the body heatings $\underset{\alpha}{s}$, which do occur in the form (5.21) of the principle of dissipation in view of (5.22), have here been eliminated. The principles (5.12) of balance of mass, linear momentum, rotational momentum, and energy make it possible to eliminate any one $\underset{\alpha}{\overset{+}{c}}$, $\underset{\alpha}{\overset{+}{\mathbf{m}}}$, $\underset{\alpha}{\mathbf{T}} - \underset{\alpha}{\mathbf{T}}^{\mathsf{T}}$, and $\underset{\alpha}{\overset{+}{e}}$ from (5.27) if we wish to.

In the important special case when all constituents of the mixture have the same coldness, so that

$$\underset{\alpha}{\vartheta} = \vartheta, \qquad \alpha = 1, 2, \ldots, \mathfrak{k}, \tag{5.28}$$

the terms in $\underset{\alpha}{\overset{+}{e}}$ drop out of (5.27) because of $(5.12)_4$, so that the principle of dissipation now does not restrict the growths of energy. The relations (5.22) reduce to

$$\begin{aligned} \phi &= \vartheta \mathbf{h}_{\mathrm{I}} - \underset{\alpha}{\rho}\underset{\alpha}{\eta}\underset{\alpha}{\mathbf{u}}, \\ \sigma &= \vartheta s_{\mathrm{I}}. \end{aligned} \tag{5.29}$$

The second of these shows that if (5.20) is the correct axiom of dissipation for mixtures, then the working of the body forces against diffusion, which does indeed furnish body heating according to $(5.14)_5$, does not contribute to the growth of caloric†. In the special case when the body forces $\underset{\alpha}{\mathbf{b}}$ are all the same, $s = s_{\mathrm{I}}$, and the distinction just made becomes vacuous.

Turning to $(5.29)_1$, we see that it may be written in the form

$$\phi = \vartheta(\mathbf{h} + \Sigma \underset{\alpha}{\mathbf{C}}^{\mathsf{T}} \underset{\alpha}{\mathbf{u}}), \tag{5.30}$$

where

$$\underset{\alpha}{\mathbf{C}} = -\underset{\alpha}{\mathbf{T}} + \underset{\alpha}{\rho}(\underset{\alpha}{\psi} + \tfrac{1}{2}\underset{\alpha}{u}^2)\mathbf{1}. \tag{5.31}$$

For a mixture of perfect fluids, since $\underset{\alpha}{\mathbf{T}} = -\underset{\alpha}{p}\mathbf{1}$, where p is the partial pressure of the αth constituent, (5.30) becomes

$$\underset{\alpha}{\mathbf{C}} = \underset{\alpha}{\rho}(\underset{\alpha}{p}/\underset{\alpha}{\rho} + \underset{\alpha}{\psi} + \tfrac{1}{2}\underset{\alpha}{u}^2)\mathbf{1}. \tag{5.32}$$

As I showed long ago‡, in a dilute mixture of perfect fluids having caloric

† This result is consistent with MAXWELL's kinetic theory (to within the usual identifications) and the continuum theories of GREEN & NAGHDI and MÜLLER but inconsistent with the claim of ERINGEN & INGRAM, Equalities (8.8) and (8.10) of their paper cited below.

‡ For example, see Section 255 of *The Classical Field Theories*, *op. cit.* For the case of a mixture of ideal gases in equilibrium, this result may be verified from well known explicit formulae for $\underset{\alpha}{p}$ and $\underset{\alpha}{\psi}$.

equations of state, the chemical potential $\underset{\mathfrak{a}}{\mu}$ of the $\mathfrak{a}^{\text{th}}$ constituent is given by $\underset{\mathfrak{a}}{\mu} = \underset{\mathfrak{a}}{\psi} + \underset{\mathfrak{a}}{p}/\underset{\mathfrak{a}}{\rho}$. By (5.32) we see that $\underset{\mathfrak{a}}{\mathbf{C}}$ for a mixture of such fluids is $\underset{\mathfrak{a}}{\rho}(\underset{\mathfrak{a}}{\mu} + \frac{1}{2}\underset{\mathfrak{a}}{u}^2)\mathbf{1}$. More generally, $\underset{\mathfrak{a}}{\mathbf{C}}$ as given by (5.31) may be called the *chemical potential tensor*. By (5.30), the action of this tensor upon the diffusion velocity provides a flux of caloric per unit coldness. In the special case when the chemical potential of the $\mathfrak{a}^{\text{th}}$ constituent is proportional to the concentration of that constituent, so that $\underset{\mathfrak{a}}{\mathbf{C}} = \underset{\mathfrak{a}}{c}\mathbf{K}$, say, then the sum in (5.29) vanishes, and $\boldsymbol{\phi} = \vartheta\mathbf{h}$ as in the case of a single medium. This result holds, in particular, in a dilute mixture of perfect fluids such that $\underset{\mathfrak{a}}{\mu}$ is independent of $\mathfrak{a}$ and $\underset{\mathfrak{a}}{\mathbf{u}} = \mathbf{0}$. In the general case, so long as the constituents have all the same temperature, (5.30) shows how the concept of chemical potential enters the theory naturally, without recourse to thermostatics or local equations of state. It is simply the peculiar pressure tensor, augmented by a hydrostatic pressure equal to the sum of the peculiar free energetic and the kinetic energy of diffusion for the constituent.

If we now go back to $(5.14)_4$, we may write the heating flux $\mathbf{h}$ also in terms of the chemical potential tensor $\underset{\mathfrak{a}}{\mathbf{C}}$, namely,

$$\mathbf{h} = \mathbf{h}_{\mathrm{I}} - \Sigma(\underset{\mathfrak{a}}{\mathbf{C}}^{\mathrm{T}} + \theta\underset{\mathfrak{a}\mathfrak{a}}{\rho\eta}\mathbf{1})\underset{\mathfrak{a}}{\mathbf{u}}. \tag{5.33}$$

Special cases of this formula, showing how the chemical potential and the caloric of a constituent contribute to the heating flux, appear in the literature of the linear "irreversible thermodynamics", as will be remarked below.

From (5.29) we see that not only net heating flux but also diffusion may give rise to caloric flux. Indeed, in a vessel containing two nonconducting fluids at rest in mean, at uniform temperature, and insulated from the exterior, irreversible diffusion may occur, and in this process the calory will increase. This possibility is covered by $(5.29)_1$, in which we see that even if $\mathbf{h}_{\mathrm{I}} = \mathbf{0}$, the resultant rate of increase div $\boldsymbol{\phi}$ will not generally vanish if diffusion is occurring.

Returning now to the general case, when each constituent may have its own peculiar coldness $\underset{\mathfrak{a}}{\vartheta}$, we can introduce a *thermokinetic process* for the mixture, namely, the set of fields

$$\underset{\mathfrak{b}}{\mathbf{x}}(\underset{\mathfrak{b}}{\mathbf{X}}, t), \qquad \underset{\mathfrak{b}}{\vartheta}(\mathbf{x}, t), \qquad \mathfrak{b} = 1, 2, \ldots, \mathfrak{k}, \tag{5.34}$$

the former defined for all $\underset{\mathfrak{b}}{\mathbf{X}}$ in the body $\underset{\mathfrak{b}}{\mathcal{B}}$, the latter for all $\mathbf{x}$ delivered by the former through (5.1), and both when $-\infty < t < k$ for some k. For a region of space in which a given thermokinetic process is occurring†, a *calorodynamic process* is a set of fields of two kinds. First, those occurring in the theory of a single body (*cf.* (2.55)):

$$\underset{\mathfrak{a}}{\mathbf{T}}(\mathbf{x}, t), \quad \underset{\mathfrak{a}}{\mathbf{h}}(\mathbf{x}, t), \quad \underset{\mathfrak{a}}{\varepsilon}(\mathbf{x}, t), \quad \underset{\mathfrak{a}}{\eta}(\mathbf{x}, t), \quad \underset{\mathfrak{a}}{\mathbf{b}}(\mathbf{x}, t), \quad \underset{\mathfrak{a}}{s}(\mathbf{x}, t), \tag{5.35}$$

and, second, the several growths describing the reactions and exchanges in the mixture:

$$\underset{\mathfrak{a}}{\overset{+}{c}}(\mathbf{x}, t), \quad \underset{\mathfrak{a}}{\overset{+}{\mathbf{m}}}(\mathbf{x}, t), \quad \underset{\mathfrak{a}}{\overset{+}{\mathbf{M}}}(\mathbf{x}, t), \quad \underset{\mathfrak{a}}{\overset{+}{e}}(\mathbf{x}, t), \quad \underset{\mathfrak{a}}{\overset{+}{\eta}}(\mathbf{x}, t), \tag{5.36}$$

provided these be subject to the *general axioms of the thermodynamics of mixtures*, namely, (5.11), (5.12), (5.18), and (5.20). Thus to express the ***constitutive axiom***‡ for the thermodynamic theory of diffusing, reacting mixtures we may use just the same words as for a single body (Lecture 2): *For given bodies $\underset{\mathfrak{a}}{\mathcal{B}}$ at time t, the history of the thermokinetic process to which neighborhoods of $\underset{\mathfrak{a}}{\mathbf{X}}$ have been subject determines a calorodynamic process for the bodies $\underset{\mathfrak{a}}{\mathcal{B}}$ at the particles* $\underset{\mathfrak{a}}{\mathbf{X}}$. Further, of course, the constitutive relations must be frame-indifferent, just as for a single body§.

† That is, a region of space occupied simultaneously by a region of each body $\underset{\mathfrak{a}}{\mathcal{B}}$.

‡ This axiom seems to include all the proposals made so far. The idea that the growths (5.36), more or less, should join the quantities (5.35) as dependent variables seems to have arisen in part from my first notes on mixtures (1957) and from Section 295 of *The Classical Field Theories, op. cit.,* and later work suggested by the ideas there. *Cf.* also the lines following (130.5) in *The Non-Linear Field Theories of Mechanics,* in FLÜGGE's *Handbuch der Physik,* vol. III/3, Springer-Verlag, 1965.

§ The general principles governing constitutive equations are:

1. Determinism
2. Equipresence
3. Local action
4. Frame-indifference
5. Dissipation

For a simple discussion of these in the case of a single body, see Lecture 1 in *Six Lectures on Modern Natural Philosophy,* Springer-Verlag, New York, 1966. The constitutive axiom given above is so phrased as to satisfy the principles of determinism, equipresence, local action, and dissipation. To implement the principle of frame-indifference, one must assume that the quantities $\underset{\mathfrak{a}}{\rho}$, $\underset{\mathfrak{a}}{\mathbf{T}}$, $\underset{\mathfrak{a}}{\mathbf{b}}$, $\underset{\mathfrak{a}}{\mathbf{h}}$, $\underset{\mathfrak{a}}{\varepsilon}$, $\underset{\mathfrak{a}}{\theta}$, η, and $\underset{\mathfrak{a}}{s}$ are frame-indifferent, as for a single medium. From this assumption and the definitions of $\underset{\mathfrak{a}}{\dot{\mathbf{x}}}$ and $\underset{\mathfrak{a}}{\ddot{\mathbf{x}}}$ follow definite rules of transformation under change of frame for the growths $\underset{\mathfrak{a}}{\overset{+}{c}}$, $\underset{\mathfrak{a}}{\overset{+}{\mathbf{m}}}$, $\underset{\mathfrak{a}}{\overset{+}{\mathbf{M}}}$, $\underset{\mathfrak{a}}{\overset{+}{e}}$, and $\underset{\mathfrak{a}}{\overset{+}{\eta}}$. In these considerations $\underset{\mathfrak{a}}{\ddot{\mathbf{x}}}$ is that frame-indifferent field which reduces to the peculiar acceleration in an inertial frame.

Like its special case for a single body in Lecture 2, this general constitutive axiom can be reduced to an ***effective principle of thermodynamic determinism,*** namely: *For each* $\mathfrak{a}$, *the seven fields*

$$\underset{\mathfrak{a}}{\mathbf{T}}, \quad \underset{\mathfrak{a}}{\mathbf{h}}, \quad \underset{\mathfrak{a}}{\psi}, \quad \underset{\mathfrak{a}}{\eta}, \quad \underset{\mathfrak{a}}{\overset{+}{c}}, \quad \underset{\mathfrak{a}}{\overset{+}{\mathbf{m}}}, \quad \underset{\mathfrak{a}}{\overset{+}{e}} \tag{5.37}$$

are determined at $\underset{\mathfrak{a}}{\mathbf{X}}$ *by the history of an arbitrary thermokinetic process in a neighborhood of each* $\underset{\mathfrak{b}}{\mathbf{X}}$ *presently occupying the same place as* $\underset{\mathfrak{a}}{\mathbf{X}}$, *in such a way that the sums of the last three are zero, that the sum of the first is a symmetric tensor field, and that the reduced dissipation inequality* (5.27) *is satisfied identically.*

While many constitutive theories of mixtures have been proposed in the last 100 years, only one of these seems to be satisfactory†, namely, the

† Only three others rest on modern thermodynamic principles:

A. E. Green & P. M. Naghdi, "A dynamical theory of interacting continua," *Intern. J. Eng. Sci.*, **3**:231–241 (1965); "A theory of mixtures," *Arch. Rational Mech. Anal.*, **24**:243–263 (1967); and later papers concerning applications and generalizations.

A. C. Eringen & J. D. Ingram, "A continuum theory of chemically reacting media–I," *Intern. J. Eng. Sci.*, **3**:197–212 (1965), and J. D. Ingram & A. C. Eringen, "A continuum theory of chemically reacting media–II, Constitutive equations of reacting fluid mixtures," *Intern. J. Eng. Sci.*, **5**:289–322 (1967).

R. M. Bowen, "Toward a thermodynamics and mechanics of mixtures," *Arch. Rational Mech. Anal.*, **24**:370–403 (1967).

These three theories are all deficient in at least one of the following respects:

1. Failure to fall in equilibrium into agreement with classical thermochemistry.
2. Failure to follow the lead of Maxwell's kinetic theory of gas mixtures and the later basic phenomenological analyses and so recognize that
   A. Each constituent has its own stress, flux of energy, *etc.*, with corresponding equations of growth.
   B. There is but a single dissipation inequality for the mixture as a whole, not individual ones for the several constituents.
3. Failure to be general enough to include anything beyond ideal mixtures.
4. Failure to adopt the principle of equipresence.

Note added in March, 1969. While my manuscript originally contained details substantiating the foregoing summary, I have decided to omit them, first because most of the criticisms have been published meanwhile by others, and, second, because a considerable measure of agreement seems now to have been reached by the major authors. To the extent I can follow the notations and concepts in a new manuscript of theirs which Green & Naghdi have kindly let me see, I judge that they have reconciled themselves to my basic equations of 1957 (in form though not, they say, in motivation), to the dissipation inequality (5.20) proposed by Bowen and me, and, theory of MÜLLER.† MÜLLER adopts all the general principles set forth above, including the principle of equipresence, and his theory is consistent in appropriate circumstances with all the classical results of thermochemistry. He considers only the case when $\underset{\alpha}{\overset{+}{c}} = 0$ and $\underset{\alpha}{\overset{+}{\mathbf{M}}} = \mathbf{0}$, excluding the possibility of chemical reactions‡ and exchanges of rotational momentum, and when all constituents have a common coldness§. $\underset{\alpha}{\vartheta} = \vartheta$ for all α. In this case, as we have remarked already, the growths of energy $\underset{\alpha}{\overset{+}{e}}$ drop out of the reduced dissipation inequality (5.27). Not only that, we may also show that that inequality can be reduced to a form in which the quantities $\underset{\alpha}{\eta}$, $\underset{\alpha}{\psi}$, and $\underset{\alpha}{\mathbf{h}}$ occur only in the combinations which define η, ψ, $\mathbf{h}$, and $\boldsymbol{\phi}$. This same fact may be verified also directly on the basis of (5.21), $(5.29)_2$, (2.18), and $(5.11)_3$ alone, without using $(5.11)_5$ or $(5.29)_1$. Thus it is possible to construct a theory without commitment to any particular relation between the flux of energy $\mathbf{h}$ and the flux of caloric $\boldsymbol{\phi}$. This is the course adopted by MÜLLER, extending the idea he had used earlier for a single medium¶. Although MÜLLER makes no use of $(5.29)_1$, he holds fast to $(5.29)_2$, thus selecting as the temperature that associated with the body heating rather than, except possibly by later implication, that associated with the contact heating.

In principle MÜLLER considers a mixture defined by constitutive equations in which the dependent variables

$$\psi, \quad \eta, \quad \mathbf{h}, \quad \boldsymbol{\phi}, \quad \underset{\alpha}{\mathbf{T}}, \quad \underset{\alpha}{\overset{+}{\mathbf{m}}}, \tag{5.38}$$

---

tacitly, to the principle of equipresence. BOWEN, on the other hand, has seen how to interpret his general framework as a special case of mine, as he explains in a new paper written jointly with J. C. WIESE, "Diffusion in mixtures of elastic materials," *Internat. J. Eng. Sci.*, in press.

† I. MÜLLER, "A thermodynamic theory of mixtures of fluids," *Arch. Rational Mech. Anal.*, **28**:1–39 (1968).

‡ However, some of the preliminary apparatus, including the basic dissipation inequality numbered (3.9) in his paper and the reduction for frame-indifference, are valid more generally.

§ In a later study carried out in collaboration with N. T. DUNWOODY, MÜLLER considers chemical reactions and also allows the coldnesses $\underset{\alpha}{\vartheta}$ to differ from one another. Thus he is able to discuss the manner in which the constituents of a reacting mixture may tend to assume a common temperature as the flow proceeds: $\underset{\alpha}{\vartheta} \to$ a common limit ϑ as $t \to \infty$. See the paper, "A thermodynamic theory of two chemically reacting ideal gases with different temperatures," *Arch. Rational Mech. Anal.*, **29**:344–369 (1968).

¶ I. MÜLLER, "On the entropy inequality," *Arch. Rational Mech. Anal.*, **26**:118–141 (1967).

$\underset{\mathfrak{a}}{\Sigma}\mathbf{T}$ being symmetric, are taken as functions of

$$\underset{\mathfrak{b}}{\rho}, \qquad \operatorname{grad} \underset{\mathfrak{b}}{\rho}, \qquad \underset{\mathfrak{b}}{\dot{\mathbf{x}}}, \qquad \operatorname{grad} \underset{\mathfrak{b}}{\dot{\mathbf{x}}}, \qquad \theta, \qquad \operatorname{grad} \theta, \qquad \mathfrak{b} = \mathit{1}, \mathit{2}, \ldots, \mathfrak{k}, \tag{5.39}$$

subject to the requirements of material frame-indifference and dissipation. As has been explained above, the variables describing energy and calory, both dependent and independent, refer only to the mixture as a whole, not to its constituents, since a common temperature θ reigns at each place and time. Second, since the deformation gradients of the constituents are allowed to enter the constitutive equations only through the densities $\underset{\mathfrak{b}}{\rho}$ and their gradients, the constituents are *fluids* in the sense† defined by NOLL. From this fact and the principle of indifference it can be shown in the now common way that all the response functions are isotropic functions. Third, the density gradients grad $\underset{\mathfrak{b}}{\rho}$ are included as independent variables specifically so as to avoid the difficult choice between, on the one hand, a theory too special for consistency with classical thermochemistry in all cases or, on the other, an artificial framework not motivated by the classical special theories of mixtures. Indeed, MÜLLER's theory was put forward especially so as to resolve the dilemma offered by earlier attempts.

So as to obtain explicit results, MÜLLER restricts his analysis to a binary mixture: $\mathfrak{k} = \mathit{2}$. While using only essentially formal mathematics, his argument is long from the number of variables and effects of interaction they represent. Certain results of the expected kind he is able to prove immediately as consequences‡ of the basic dissipation inequality (5.21), reduced as usual by elimination of the peculiar body forces and body heatings. First, he shows that

$$\psi \text{ is independent of the gradients of the velocities and the temperature gradient;} \tag{5.40}$$

second, that

$$\psi - \tfrac{1}{2}\underset{\mathfrak{a}}{\Sigma}\underset{\mathfrak{a}}{c}u^2 = \psi_{\mathrm{I}}(\theta, \underset{\mathfrak{b}}{\rho}, \operatorname{grad} \underset{\mathfrak{c}}{\rho} \cdot \operatorname{grad} \underset{\mathfrak{b}}{\rho}). \tag{5.41}$$

Moreover, the classical formula for calculating the caloric from the free

† Of course, not precisely, since NOLL did not treat mixtures at all, but a definition of fluids in a theory of mixtures of materials of the second grade along the lines parallel to those in NOLL's theory of single simple materials can be framed.

‡ Since this theory is one of viscous fluids, the results cannot simply be read off from the general formulation of GURTIN (Lecture 3), although, as GURTIN remarks, his procedure can be generalized insofar as only Axioms 1, 2, and 3 are needed.

energetic remains valid:

$$\eta = -\partial_\theta \psi = -\partial_\theta \psi_{\mathrm{I}}. \tag{5.42}$$

Thus the caloric does not depend on the kinetic energy of diffusion.

While the rather general relations derived by MÜLLER are too complicated to present in this lecture, some simple and explicit results emerge for equilibrium, here defined as a condition in which all constituents move at a common, uniform velocity with a common, uniform temperature, and the density of each constituent is uniform and steady. Denoting results appropriate to equilibrium by a subscript 0, MÜLLER shows that, among other things,

$$\underset{\mathfrak{a}}{\mathbf{T}}_0 = -\underset{\mathfrak{a}}{p}\mathbf{1}, \tag{5.43}$$

where

$$\underset{\mathfrak{a}}{p} = \rho\underset{\mathfrak{a}}{\rho}(\partial_{\underset{\mathfrak{a}}{\rho}}\psi_{\mathrm{I}}) \pm \theta[\partial_{\mathbf{v}^{12}}(\boldsymbol{\phi} - \vartheta\mathbf{h} - \vartheta\Sigma\underset{\mathfrak{a}}{\mathbf{T}}^{\mathsf{T}}\underset{\mathfrak{a}}{\mathbf{u}})]_0, \qquad \mathfrak{a} = 1, 2, \tag{5.44}$$

$\mathbf{v}^{12}$ being defined as the velocity of the first constituent relative to the second: $\mathbf{v}^{12} \equiv \underset{1}{\dot{\mathbf{x}}} - \underset{2}{\dot{\mathbf{x}}}$. Hence

$$p \equiv \Sigma\underset{\mathfrak{a}}{p} = \rho^2\partial_\rho\hat{\psi}_{\mathrm{I}}, \tag{5.45}$$

where $\hat{\psi}_{\mathrm{I}}(\theta, \rho, \underset{1}{c}) \equiv \psi_{\mathrm{I}}(\theta, \underset{1}{\rho}, \underset{2}{\rho}, 0)$. From these results consistency of MÜLLER's theory with classical thermochemistry can be established in those cases when there is a basis for comparison. Moreover, MÜLLER shows that

$$[\partial_{\mathbf{v}^{12}}(\boldsymbol{\phi} - \vartheta\mathbf{h})]_0 = \vartheta\underset{1}{\rho}\underset{2}{\rho}(\partial_{\underset{1}{\rho}}\psi_{\mathrm{I}} - \partial_{\underset{2}{\rho}}\psi_{\mathrm{I}})_0\mathbf{1}. \tag{5.46}$$

Thus even in linear approximation we shall not have $\boldsymbol{\phi} = \vartheta\mathbf{h}$ if diffusion is taking place. In physical chemistry the partial derivatives occurring on the right-hand side of (5.46) are called chemical potentials, and the linear "irreversible thermodynamics" postulates that (5.46) holds in general, not merely in equilibrium.

MÜLLER then takes up in detail the case when the constitutive functions are linear. He obtains and interprets fully explicit forms for all the equations, inequalities restricting the signs of the coefficients, and specializations to important cases, one of which deserves remark now.

If the density gradients $\underset{\mathfrak{b}}{\boldsymbol{\rho}}$ are omitted from the list (5.39) of independent variables in the constitutive relations, the theory is said to describe a *simple mixture*. For such a mixture, MÜLLER proves from the dissipation inequality that the peculiar free energetic of constituent $\mathfrak{a}$ at a given temperature is determined by the density of constituent $\mathfrak{a}$ alone:

$$\underset{\mathfrak{a}}{\psi} = \underset{\mathfrak{a}}{\psi}(\underset{\mathfrak{a}}{\rho}, \theta). \tag{5.47}$$

In this way the free energy of the mixture is the sum of the free energies of each constituent, calculated as if the other were not present. Many of the older theories, either expressly or tacitly, assumed such to be the case, and it has been a primary aim of recent theories to avoid a formulation so restrictive as somehow to imply (5.47) in all cases†. In such a mixture (5.44) becomes‡

$$\underset{\mathfrak{a}}{p} = \underset{\mathfrak{a}}{\rho}^2 \partial_{\underset{\mathfrak{a}}{\rho}} \underset{\mathfrak{a}}{\psi}. \tag{5.48}$$

The earlier authors, at least mainly, did not include density gradients in the list (5.39) of independent variables. The classical theories and most of the more recent theories of a single fluid, being special cases of the theory of simple materials, enjoyed great success without needing to consider the added complication of effects dependent on grad ρ, so as a first step in the theory of mixtures it seemed enough to leave them out there, too. Here what seemed enough was in fact deficient. MÜLLER's theory, also, reduces in the special case of a single fluid, when $\mathfrak{k} = 1$, to a theory in which density gradients drop out of the list (5.39); in linear approximation, it further reduces to just the classical theory, not to some more complicated one. The occurrence of density gradients in the list (5.39), far from indicating a desire to be general for generality's sake, seems necessary in order to get a theory of equilibrium that is merely as general as classical thermochemistry. If density gradients are omitted from the start, one is forced back to the unduly special property expressed by (5.47). In particular, the results when grad $\underset{\mathfrak{b}}{\rho} = \mathbf{0}$ is imposed as a specializing condition are not the same as those when grad $\underset{\mathfrak{b}}{\rho}$ is just left out of account§.

MÜLLER considers also ideal mixtures as defined in thermochemistry, and mixtures of ideal gases, which are special cases of them, corresponding to particular functional forms for the $\underset{\mathfrak{a}}{\psi}$ that in effect express DALTON's law of partial pressures.

† The restricted scope of (5.47) is clear even within the special framework of my note of 1957. *Cf.* Section 255 of *The Classical Field Theories*, *op. cit.*

‡ In fact, something has to be assumed here about the behavior of $\underset{\mathfrak{a}}{p}$ and $\underset{\mathfrak{a}}{\psi}$ as $\underset{\mathfrak{a}}{\rho} \to 0$.

§ The results gotten from thermodynamics based on the Clausius-Duhem inequality are very sensitive to the constitutive class admitted from the start. I do not see that this is a defect. If, in any theory, not necessarily thermodynamics, we lay down conditions on $f(x, y)$ and then show that in fact the dependence on y is illusory, we cannot expect to get only the same results as if we had assumed that $f(x, y) = g(x)$ from the start.

To go into further detail regarding MÜLLER's theory would take more time than is allowed here and would needlessly duplicate the text of his memoir†. The outline already given satisfies the purpose of this lecture: to show that COLEMAN & NOLL's method, applied to my original framework for theories of mixtures and supplemented by the principle of equipresence, suffices to obtain a rational and explicit thermodynamics of diffusion, incorporating all results in the accepted theory of chemical equilibrium.

Bibliographical Note

Since MÜLLER in his paper cites the manuscript of my lecture on this subject delivered at the Drexel Institute of Technology on January 19, 1967, I have to explain that that lecture was roughly the same as this one only through about (5.16). The rest of the lecture as printed here has been rewritten entirely after I had the opportunity of studying MÜLLER's work. In particular, my formulation of the caloric inequality for mixtures corrects the unpublished one of 1967.

† *Note added in March, 1969.* MÜLLER's theory has been analyzed further by M. L. DORIA, "Some general results for non-reacting binary mixtures of fluids," *Arch. Rational Mech. Anal.*, **32**:343–368 (1969). DORIA obtains and interprets conditions which are not only necessary but also sufficient for MÜLLER's constitutive relations to satisfy the reduced dissipation inequality.

LECTURE 6

THERMODYNAMICS OF CHEMICAL REACTIONS

General relations for exchanges of mass among the constituents of a mixture are included in the framework set up in the preceding lecture. Specifically, let $\underset{\mathfrak{a}}{\rho}$ be the peculiar density of constituent $\mathfrak{a}$, where $\mathfrak{a} = 1, 2, \ldots, \mathfrak{k}$. Then the density of the mixture is defined by

$$\rho \equiv \sum_{\mathfrak{a}=1}^{\mathfrak{k}} \underset{\mathfrak{a}}{\rho}. \tag{5.4}$$

The *mass fraction* $\underset{\mathfrak{a}}{c}$ of the constituent $\mathfrak{a}$ is given by

$$\underset{\mathfrak{a}}{c} \equiv \frac{\underset{\mathfrak{a}}{\rho}}{\rho}, \tag{5.5}$$

so that

$$\sum_{\mathfrak{a}=1}^{\mathfrak{k}} \underset{\mathfrak{a}}{c} = 1. \tag{5.6}$$

Let $\underset{\mathfrak{a}}{\dot{\mathbf{x}}}$ be the peculiar velocity of constituent $\mathfrak{a}$. The *mass growth* $\underset{\mathfrak{a}}{\overset{+}{c}}$ of constituent $\mathfrak{a}$, defined by $(5.11)_1$, may be expressed directly in terms of the mean motion and the diffusion velocity $\underset{\mathfrak{a}}{\mathbf{u}}$. Indeed, by using (2.3) we easily show that

$$\rho \underset{\mathfrak{a}}{\dot{c}} = -\operatorname{div}(\underset{\mathfrak{a}}{\rho} \underset{\mathfrak{a}}{\mathbf{u}}) + \rho \underset{\mathfrak{a}}{\overset{+}{c}}, \tag{6.1}$$

where the dot denotes the time derivative following the velocity $\dot{\mathbf{x}}$ of the mixture. The condition expressing the conservation of total mass in the mixture is

$$\sum_{\mathfrak{a}=1}^{\mathfrak{k}} \underset{\mathfrak{a}}{\overset{+}{c}} = 0. \tag{5.12$_1$}$$

In ordinary chemistry, however, the exchanges of mass are restricted to a special kind, according to the laws of "chemical reactions". In such reactions, the constituents combine and dissociate only in definite proportions. These proportions are commonly explained by saying that each substance consists of "molecules"; that each molecule is composed of a certain whole number of "atoms" of a few specified kinds. These atoms have fixed "atomic weights", and the "molecular weight" of a substance is the sum of the weights of the atoms in one molecule of that substance. In the reactions, molecules are created or destroyed, but the atoms are permanent in number and nature. The terms "atom" and "molecule" are merely convenient for visualizing the rules of definite combination and need not be thought of as denoting corpuscles, nor need we limit the interpretation to strictly "chemical" changes. All that needs be assumed is that each constituent is made up by combination in fixed proportions from certain individually indestructible constituents. Nevertheless, the terms "atomic" and "molecular" are so familiar that I shall occasionally use them in describing the pure phenomenology of reactions.

The theory of this special kind of exchange of mass is called *chemical kinetics*. While it has a long history in the last century and has been developed in this one by JOUGUET, DE DONDER, and ARIS, the first really compact yet complete and invariant presentation of it has been effected by BOWEN†.

Let there be given κ atomic substances with atomic weights a^{α}, $\alpha = 1, 2, \ldots, \kappa$. These are supposed capable of combining so as to form $\mathfrak{f}$ molecular substances composed from them in definite proportions. Thus the molecular substances have molecular weights $m^{\mathfrak{b}}$ given by

$$m^{\mathfrak{b}} = \sum_{\alpha=1}^{\kappa} t_{\alpha}{}^{\mathfrak{b}} a^{\alpha}, \tag{6.2}$$

where $t_{\alpha}{}^{\mathfrak{b}}$ is a non-negative integer, to be interpreted as the number of moles of the αth atomic substance in one mole of the $\mathfrak{a}^{\text{th}}$ molecular substance. In chemical kinetics the numbers a^{α} and $t_{\alpha}{}^{\mathfrak{b}}$ are supposed given *a priori*, and the basic axiom asserts the *permanence of atomic substances:*

$$\sum_{\mathfrak{a}=1}^{\mathfrak{f}} \frac{t_{\alpha}{}^{\mathfrak{a}}}{m^{\mathfrak{a}}} \overset{+}{c}_{\mathfrak{a}} = 0, \qquad \alpha = 1, 2, \ldots, \kappa. \tag{6.3}$$

It is preferable to eliminate any dependent columns of the matrix $\|t_{\alpha}{}^{\mathfrak{a}}\|$ and write (6.2) in the form

$$m^{\mathfrak{a}} = \sum_{\alpha=1}^{\nu} s_{\alpha}{}^{\mathfrak{a}} \overline{m}^{\alpha}, \tag{6.4}$$

where the ν molecular weights $\overline{m}^{\alpha}$ are certain linear combinations of the $m^{\mathfrak{a}}$, and $\operatorname{rank}\|s^{\mathfrak{a}}{}_{\alpha}\| = \nu = \operatorname{rank}\|t_{\alpha}{}^{\mathfrak{a}}\|$. It is then possible to write the axiom (6.3) in the form‡

$$\sum_{\mathfrak{a}=1}^{\mathfrak{f}} \frac{s_{\alpha}{}^{\mathfrak{a}}}{m^{\mathfrak{a}}} \overset{+}{c}_{\mathfrak{a}} = 0, \qquad \alpha = 1, 2, \ldots, \kappa. \tag{6.5}$$

† R. M. BOWEN, "On the stoichiometry of chemically reacting mixtures," *Arch. Rational Mech. Anal.*, **29**:114–124 (1968).

‡ *Cf.* Section 159A of *The Classical Field Theories*, in FLÜGGE's *Handbuch der Physik*, vol. III/1, Springer-Verlag, 1960.

We shall now express and interpret these equations in terms of vectors in a $\mathfrak{f}$-dimensional inner product space† $\mathcal{U}$, called the *constituent space*. Let $(\mathbf{e}_1, \mathbf{e}_2, \ldots, \mathbf{e}_{\mathfrak{f}})$ be a basis in $\mathcal{U}$, and let $\mathbf{e}^1, \mathbf{e}^2, \ldots, \mathbf{e}^{\mathfrak{f}}$ be its reciprocal basis. Set

$$\mathbf{m} \equiv \sum_{\mathfrak{a}=1}^{\mathfrak{f}} m^{\mathfrak{a}} \mathbf{e}_{\mathfrak{a}}, \qquad \boldsymbol{\omega} \equiv \sum_{\mathfrak{a}=1}^{\mathfrak{f}} (c_{\mathfrak{a}}/m^{\mathfrak{a}}) \mathbf{e}^{\mathfrak{a}}. \tag{6.6}$$

The component $c_{\mathfrak{a}}/m^{\mathfrak{a}}$ of $\boldsymbol{\omega}$ is called the *molar concentration* of constituent $\mathfrak{a}$, and the same name may be applied to the vector $\boldsymbol{\omega}$. The identity (5.6) takes the form

$$\mathbf{m} \cdot \boldsymbol{\omega} = 1. \tag{6.7}$$

Next, (6.1) becomes simply

$$\dot{\omega} = \boldsymbol{\sigma} + \overset{+}{\boldsymbol{\omega}}, \tag{6.8}$$

where

$$\boldsymbol{\sigma} = -\sum_{\mathfrak{a}=1}^{\mathfrak{f}} \frac{1}{\rho m^{\mathfrak{a}}} \operatorname{div} (\rho_{\mathfrak{a}} \mathbf{u}_{\mathfrak{a}}) \mathbf{e}^{\mathfrak{a}}, \qquad \overset{+}{\boldsymbol{\omega}} = \sum_{\mathfrak{a}=1}^{\mathfrak{f}} (\overset{+}{c}_{\mathfrak{a}}/m^{\mathfrak{a}}) \mathbf{e}^{\mathfrak{a}}. \tag{6.9}$$

Thus the rate of increase of the molar concentration is divided into two parts, one being the resultant effect of diffusion, the other the effect of reactions. In chemical kinetics, it is the *molar growth* $\overset{+}{\boldsymbol{\omega}}$ rather than *diffusive growth* $\boldsymbol{\sigma}$ that is the object of study.

So far, we have done no more than introduce notation. The vector $\mathbf{m}$ is fixed *a priori* in $\mathcal{U}$, and the vectors $\boldsymbol{\omega}$ and $\overset{+}{\boldsymbol{\omega}}$ may vary in any way consistent with (6.7) and (6.8), where $\boldsymbol{\sigma}$ is regarded as determined by the motions of the constituents. The axiom of ***permanence of atomic substances*** now takes an elegant and pellucid form: *The molar growth* $\overset{+}{\boldsymbol{\omega}}$ *lies in a* $(\mathfrak{f} - \nu)$*-dimensional subspace* $\mathcal{V}$ *that is normal to* $\mathbf{m}$. To see that this axiom does indeed deserve its name, we note first that $\mathcal{U}$ is the direct sum of $\mathcal{V}$ and its orthogonal complement $\mathcal{V}^{\perp}$:

$$\mathcal{U} = \mathcal{V} \oplus \mathcal{V}^{\perp}, \tag{6.10}$$

and the axiom asserts that

$$\overset{+}{\boldsymbol{\omega}} \in \mathcal{V}, \qquad \mathbf{m} \in \mathcal{V}^{\perp}, \tag{6.11}$$

† The inner product is arbitrary and without physical significance. It is introduced only for formal convenience.

where $\dim \mathcal{V}^\perp = \nu$. Let $(\mathbf{f}_1, \mathbf{f}_2, \ldots, \mathbf{f}_\nu)$ be a basis for $\mathcal{V}^\perp$. Then by $(6.11)_2$, there exist components $\bar{m}^\alpha$ such that

$$\mathbf{m} = \sum_{\alpha=1}^{\nu} \bar{m}^\alpha \mathbf{f}_\alpha, \tag{6.12}$$

and by $(6.11)_1$,

$$\overset{+}{\boldsymbol{\omega}} \cdot \mathbf{f}_\alpha = 0, \qquad \alpha = 1, 2, \ldots, \nu. \tag{6.13}$$

Finally, let $s_\alpha{}^{\mathfrak{a}}$ be the components of $\mathbf{f}_\alpha$ with respect to the basis $\mathbf{e}_{\mathfrak{a}}$:

$$\mathbf{f}_\alpha = \sum_{\mathfrak{a}=1}^{\mathfrak{f}} s_\alpha{}^{\mathfrak{a}} \mathbf{e}_{\mathfrak{a}}. \tag{6.14}$$

If we substitute (6.14) into (6.12) and compare the result with $(6.6)_1$, we recover (6.4). If we substitute $(6.9)_2$ and (6.14) into (6.13), we recover (6.5). Conversely, if we start with (6.4) and (6.5), then define $\mathbf{f}_\alpha$ in terms of its components by (6.14), we may derive (6.12) and (6.13), and hence (6.11). Thus (6.11) is equivalent to our original statement of the axiom of permanence of atomic substances.

From (6.11) it follows trivially that

$$\mathbf{m} \cdot \overset{+}{\boldsymbol{\omega}} = 0. \tag{6.15}$$

From (6.6) and $(6.9)_2$ we see that this equation is equivalent to $(5.12)_1$. That is, the axiom of permanence of atomic substances implies the conservation of total mass in the mixture. Moreover, since $\mathbf{m}$ is constant, differentiation of (6.7) yields $\mathbf{m} \cdot \dot{\boldsymbol{\omega}} = 0$. By (6.15) and (6.8) we conclude that

$$\mathbf{m} \cdot \boldsymbol{\delta} = 0. \tag{6.16}$$

This condition restricts the diffusion velocities $\underset{\mathfrak{a}}{\mathbf{u}}$ so as to be compatible with the axiom of permanence.

$\mathcal{V}$ is called the *reaction subspace*, and a *reaction* is a one-dimensional subspace $\mathcal{V}_{\mathfrak{p}}$ of $\mathcal{V}$.

When $\boldsymbol{\delta} \in \mathcal{V}$, the diffusion is *self-balancing*. In this case, by $(6.11)_1$, $\dot{\boldsymbol{\omega}} \in \mathcal{V}$. That is, when the diffusion is self-balancing, $\dot{\boldsymbol{\omega}}$ lies in the reaction subspace. This is true, in particular, when there is no diffusion at all, since then $\boldsymbol{\delta} = \mathbf{0}$. Moreover, if $\dot{\boldsymbol{\omega}} \in \mathcal{V}$, so is the time derivative of $\boldsymbol{\omega} - \boldsymbol{\omega}_0$, where $\boldsymbol{\omega}_0$ is the value of $\boldsymbol{\omega}$ at some reference time. Consequently, if $\boldsymbol{\delta} \in \mathcal{V}$, then

$$\boldsymbol{\xi} \equiv \boldsymbol{\omega} - \boldsymbol{\omega}_0 \in \mathcal{V}, \tag{6.17}$$

where of course so as to satisfy (6.7) the constant $\boldsymbol{\omega}_0$ must be such that

$$\mathbf{m} \cdot \boldsymbol{\omega}_0 = 1. \tag{6.18}$$

Any vector $\mathfrak{x}$ is called an *extent of reaction*†. The result (6.17) asserts that *when the diffusion is self-balancing, every extent of reaction must lie in the reaction subspace.*

The *null reaction* is $\{\mathbf{0}\}$. The reactions $\mathcal{V}_1, \mathcal{V}_2, \ldots, \mathcal{V}_{\mathfrak{l}}$ are *independent* if the intersection of any one with the subspace spanned by the rest is the null reaction. In view of a theorem of linear algebra, we see that *there are* $\mathfrak{k} - \nu$ *independent reactions* $\mathcal{V}_{\mathfrak{p}}$, $\mathfrak{p} = 1, 2, \ldots, \mathfrak{k} - \nu$, and

$$\mathcal{V} = \mathcal{V}_1 \oplus \mathcal{V}_2 \oplus \cdots \oplus \mathcal{V}_{\mathfrak{k}-\nu}. \tag{6.19}$$

Thus the reactions $\mathcal{V}_{\mathfrak{p}}$ are orthogonal to $\mathcal{V}^{\perp}$; they may be chosen to be mutually orthogonal as well, but since the inner product used here is a purely formal one, without intrinsic meaning, an orthogonal resolution has no advantage over any other.

If $\mathbf{x} \in \mathcal{V}$, then $\mathbf{x}$ has a unique decomposition of the form

$$\mathbf{x} = \sum_{\mathfrak{p}=1}^{\mathfrak{k}-\nu} \mathbf{x}_{\mathfrak{p}}, \qquad \mathbf{x}_{\mathfrak{p}} \in \mathcal{V}_{\mathfrak{p}}. \tag{6.20}$$

In particular, in virtue of the axiom $(6.11)_1$,

$$\overset{+}{\omega} = \sum_{\mathfrak{p}=1}^{\mathfrak{k}-\nu} \mathfrak{i}_{\mathfrak{p}}, \qquad \mathfrak{i}_{\mathfrak{p}} \in \mathcal{V}_{\mathfrak{p}}. \tag{6.21}$$

The vectors $\mathfrak{i}_{\mathfrak{p}}$, which are defined uniquely in terms of the decomposition (6.19), are the *reaction rates*.

The main theorem of chemical kinetics is the ***law of definite proportions:*** *Given the decomposition* (6.19) *of* $\mathcal{V}$ *into independent reactions, in the case when the diffusion is self-balancing an extent of reaction has a unique decomposition of the form*

$$\mathfrak{x} = \sum_{\mathfrak{p}=1}^{\mathfrak{k}-\nu} \mathfrak{x}_{\mathfrak{p}}, \qquad \mathfrak{x}_{\mathfrak{p}} \in \mathcal{V}_{\mathfrak{p}}. \tag{6.22}$$

That is, once we select a particular set of independent reactions, the extent of reaction is uniquely expressed as a sum of elements, one from each reaction. This famous law of chemical kinetics is no more than a simple consequence of linear algebra.

If we differentiate (6.22) and note (6.17), we find that

$$\dot{\omega} = \sum_{\mathfrak{p}=1}^{\mathfrak{k}-\nu} \dot{\mathfrak{x}}_{\mathfrak{p}}. \tag{6.23}$$

Comparison with (6.8) and (6.21) shows that

$$\eth = \sum_{\mathfrak{p}=1}^{\mathfrak{k}-\nu} (\dot{\mathfrak{x}}_{\mathfrak{p}} - \mathfrak{i}_{\mathfrak{p}}). \tag{6.24}$$

† Expressed in terms of components, this concept was introduced by De Donder.

Hence $\boldsymbol{\sigma} \in \mathcal{V}$. That is, the law of definite proportions is a necessary as well as sufficient condition for self-balancing diffusion. If $\boldsymbol{\sigma} = \mathbf{0}$, by (6.24) and the independence of the reactions we see that

$$\boldsymbol{\iota}_{\mathfrak{p}} = \dot{\boldsymbol{\xi}}_{\mathfrak{p}}, \qquad \mathfrak{p} = 1, 2, \ldots, \mathfrak{k} - \nu. \tag{6.25}$$

That is, in a non-diffusing mixture *the reaction rate is the time-rate of change of any corresponding extent of reaction.*

In chemistry, these relations are always expressed in terms of components corresponding to a particular choice of independent reactions. In particular, a basis $(\mathbf{g}^1, \mathbf{g}^2, \ldots, \mathbf{g}^{\mathfrak{k}-\nu})$ is selected for the reaction subspace $\mathcal{V}$, and the independent reaction $\mathcal{V}_{\mathfrak{p}}$ is taken as the subspace generated by $\mathbf{g}^{\mathfrak{p}}$. The reaction rate $\boldsymbol{\iota}_{\mathfrak{p}}$ then has the representation $\boldsymbol{\iota}_{\mathfrak{p}} = \iota_{\mathfrak{p}}\mathbf{g}^{\mathfrak{p}}$, and

$$\overset{+}{\omega}_{\mathfrak{a}} = \sum_{\mathfrak{p}=1}^{\mathfrak{k}-\nu} P_{\mathfrak{a}}{}^{\mathfrak{p}} \iota_{\mathfrak{p}}, \tag{6.26}$$

where

$$P_{\mathfrak{a}}{}^{\mathfrak{p}} = \mathbf{e}_{\mathfrak{a}} \cdot \mathbf{g}^{\mathfrak{p}}. \tag{6.27}$$

The matrix $\|P_{\mathfrak{a}}{}^{\mathfrak{p}}\|$ is called the *stoicheiometric matrix.* For details and further developments, Bowen's paper should be consulted, but we shall not need here anything beyond the abstract and general structure.

In a major memoir, Bowen† has constructed the first theory of chemical reactions on the basis of modern thermodynamics. He considers the case when no diffusion is taking place, so all the constituents partake in a common motion. For such a mixture, a *thermokinetic* process and a *calorodynamic process* are only slightly more general than those discussed in Lecture 2, being very special cases of those defined by (5.35) and (5.36) in Lecture 5. Indeed, one has to add only the extent of reaction $\boldsymbol{\xi}$ to the list (2.54), the molar growth $\overset{+}{\boldsymbol{\omega}}$ to the list (2.55). In specifying the conditions satisfied by a calorodynamic process, to those listed in Lecture 2 we need only add the axiom of permanence of atomic substances in the form

$$\overset{+}{\boldsymbol{\omega}} = \dot{\boldsymbol{\xi}}, \tag{6.28}$$

an immediate consequence of (6.8) and (6.17) when $\boldsymbol{\sigma} = \mathbf{0}$. In particular, the reduced dissipation inequality retains the classical form (2.56), and

† R. M. Bowen, "Thermochemistry of reacting materials," *J. Chem. Phys.*, **49**:1625–1637 (1968). This publication is marred by dozens of misprints. A formally similar but less general theory of dissipative processes with internal parameters had been presented earlier by B. D. Coleman & M. E. Gurtin, "Thermodynamics with internal state variables," *J. Chem. Phys.*, **47**:597–613 (1967), where similar results follow. The analysis of equilibrium, stability, and approach to equilibrium is not the same in the two studies, especially since Bowen includes $\dot{\mathbf{F}}$ as an independent variable, while Coleman & Gurtin do not.

an effective principle of determinism may be formulated in just the same way as in Lectures 2 and 5.

The particular class of constitutive relations selected for study by Bowen is:

$$\begin{aligned}\psi &= \bar{\psi}(\theta, \gamma, \mathbf{F}, \dot{\mathbf{F}}, \xi),\\ \eta &= \bar{\eta}(\theta, \gamma, \mathbf{F}, \dot{\mathbf{F}}, \xi),\\ \mathbf{T} &= \bar{\mathbf{T}}(\theta, \gamma, \mathbf{F}, \dot{\mathbf{F}}, \xi),\\ \mathbf{h} &= \bar{\mathbf{h}}(\theta, \gamma, \mathbf{F}, \dot{\mathbf{F}}, \xi),\\ \overset{+}{\omega} &= \bar{\mathbf{g}}(\theta, \gamma, \mathbf{F}, \dot{\mathbf{F}}, \xi),\end{aligned} \tag{6.29}$$

where $\bar{\psi}$, $\bar{\eta}$, . . . , $\bar{\mathbf{g}}$ are the response functions. Of course the principle of equipresence is satisfied. Since $\dot{\mathbf{F}}$ appears among the independent variables, the classical effects of viscosity are allowed for. Since $\mathbf{F}$ rather than merely ρ is included, the materials experiencing the reactions may be solids, aeolotropic or isotropic, as well as fluids. Particular kinds of material symmetry are discussed by Bowen in his paper.

Since $\dot{\mathbf{F}}$ is an argument as well as $\mathbf{F}$, we cannot blindly apply the general theorem of Gurtin presented in Lecture 3, but the method used there can be generalized, as suggested by the classic analysis of Coleman & Mizel†, who considered precisely the case when the constitutive relations are of the form (6.29) but without ξ as an independent variable and without $\overset{+}{\omega}$ as a dependent one. The domain of the response functions is assumed to be closed not only under local linear continuations but also under local quadratic continuation of $\mathbf{F}$. That is, if $\mathbf{F}$ assumes the value $\mathbf{F}(t_0)$ in some process when $t = t_0$, and if

$$\mathbf{F}_{\mathbf{U},\mathbf{V}}(t) \equiv \mathbf{F}(t_0) + (t - t_0)\mathbf{U} + \tfrac{1}{2}(t - t_0)^2\mathbf{V}, \tag{6.30}$$

we assume that at each t_0 there exists a k such that when $t_0 \leqq t < t_0 + k$, $\mathbf{F}_{\mathbf{U},\mathbf{V}}(t)$ lies in the domain of all the response functions occurring in (6.29). Since, in view of (6.28), the constitutive relation $(6.29)_5$ is a differential equation for the reaction rate ξ, in order to be able to construct a thermokinetic process passing through ξ_0 we have to assume that the initial-value problem for $(6.29)_5$ with this datum has a unique solution for some sufficiently small interval of time.

Calculating the restrictions imposed by the reduced dissipation inequality follows the now standard line and leads to some results at once. First, $\bar{\psi}$ is independent of $\dot{\mathbf{F}}$ and γ:

$$\psi = \bar{\psi}(\theta, \mathbf{F}, \xi). \tag{6.31}$$

Second, the free-energetic function $\bar{\psi}$ is a potential for the caloric:

$$\bar{\eta} = -\partial_\theta \bar{\psi}. \tag{6.32}$$

† B. D. Coleman & V. J. Mizel, "Existence of caloric equations of state in thermodynamics," *J. Chem. Phys.*, **40**:1116–1125 (1964).

Third, the reduced dissipation inequality becomes

$$\operatorname{tr}\{(\bar{\mathbf{T}}(\theta, \boldsymbol{\gamma}, \mathbf{F}, \dot{\mathbf{F}}, \boldsymbol{\xi}) - \rho\mathbf{F}\partial_{\mathbf{F}}\bar{\psi}(\theta, \mathbf{F}, \boldsymbol{\xi})^{\mathrm{T}})\dot{\mathbf{F}}\mathbf{F}^{-1}\} \\ + \rho\bar{\boldsymbol{\alpha}}(\theta, \mathbf{F}, \boldsymbol{\gamma}) * \bar{\mathbf{g}}(\theta, \boldsymbol{\gamma}, \mathbf{F}, \dot{\mathbf{F}}, \boldsymbol{\xi}) - \bar{\mathbf{h}}(\theta, \boldsymbol{\gamma}, \mathbf{F}, \dot{\mathbf{F}}, \boldsymbol{\xi}) \cdot \boldsymbol{\gamma} \geqq 0, \quad (6.33)$$

where $\boldsymbol{\alpha}$, the *chemical affinity*, is given by

$$\boldsymbol{\alpha} = \bar{\boldsymbol{\alpha}}(\theta, \mathbf{F}, \boldsymbol{\xi}) \equiv -\partial_{\boldsymbol{\xi}}\bar{\psi}(\theta, \mathbf{F}, \boldsymbol{\xi}), \quad (6.34)$$

and where "$*$" denotes the inner product in the reaction space. The sum of the first two summands on the left-hand side of (6.33) is $\rho\delta$, where δ is the internal dissipation (2.25). If the functions $\bar{\mathbf{T}}$, $\bar{\mathbf{h}}$, and $\bar{\mathbf{g}}$ are independent of $\dot{\mathbf{F}}$, then (6.33) implies that $\bar{\psi}$ is a potential also for the stress†:

$$\mathbf{T} = \bar{\mathbf{T}}(\theta, \mathbf{F}, \boldsymbol{\xi}) = \rho\mathbf{F}\partial_{\mathbf{F}}\bar{\psi}(\theta, \mathbf{F}, \boldsymbol{\xi})^{\mathrm{T}}, \quad (6.35)$$

and (6.33) takes the much simpler form (2.47) with δ given specifically by

$$\delta = \bar{\boldsymbol{\alpha}}(\theta, \mathbf{F}, \boldsymbol{\xi}) * \bar{\mathbf{g}}(\theta, \boldsymbol{\gamma}, \mathbf{F}, \boldsymbol{\xi}), \\ = \boldsymbol{\alpha} * \overset{+}{\boldsymbol{\omega}}. \quad (6.36)$$

That is, the internal dissipation is the scalar product of the chemical affinity and the molar growth. Likewise, if $\bar{\mathbf{T}}$, $\bar{\mathbf{h}}$, and $\bar{\mathbf{g}}$ are independent of $\boldsymbol{\gamma}$, then (6.33) implies that the mixture cannot conduct heat at all:

$$\bar{\mathbf{h}}(\theta, \mathbf{F}, \dot{\mathbf{F}}, \boldsymbol{\xi}) = \mathbf{0}. \quad (6.37)$$

In this case, we recover Planck's inequality (2.27), although δ remains given as a rather complicated combination of the response functions which is dependent upon $\dot{\mathbf{F}}$ as well as upon θ, $\mathbf{F}$, and $\boldsymbol{\gamma}$. If *both* $\dot{\mathbf{F}}$ and $\boldsymbol{\gamma}$ are dropped from the list of independent variables, all three results (6.35), (6.36), and (6.37) follow.

BOWEN, however, does not specialize his theory in any of these ways but rather draws conclusions from the more general inequality (6.33). His critical assumption is that the function $\bar{\mathbf{g}}$ is of such a kind that there exist θ^+, $\mathbf{F}^+$, and $\boldsymbol{\xi}^+$ satisfying

$$\bar{\mathbf{g}}(\theta^+, \mathbf{0}, \mathbf{F}^+, \mathbf{0}, \boldsymbol{\xi}^+) = \mathbf{0}. \quad (6.38)$$

That is, he assumes that at least one uniform temperature field, steady deformation field, and extent of reaction field annul the molar growth $\overset{+}{\boldsymbol{\omega}}$. From the standpoint of mechanics, an argument $(\theta^+, \mathbf{0}, \mathbf{F}^+, \mathbf{0}, \boldsymbol{\xi}^+)$ which satisfies (6.38) deserves to be called an *equilibrium state*, since it corresponds to concentrations, deformation, and extent of reaction that do not change in time at the point in question, and to a temperature field that is uniform near it. BOWEN remarks that chemists often use a more stringent definition

† This is the case in COLEMAN & GURTIN's theory of materials with internal parameters.

of equilibrium state, namely, an argument such that not only (6.38) holds but also the chemical affinity vanishes:

$$\bar{\boldsymbol{\alpha}}(\theta^+, \mathbf{F}^+, \boldsymbol{\xi}^+) = \mathbf{0}. \tag{6.39}$$

This special kind of equilibrium state may be called *strong*. An equilibrium state in which (6.39) does not hold may be called *false*†. A major aim is to find conditions that distinguish false equilibrium from strong.

Returning to the dissipation inequality (6.33), we see that by (6.38) the left-hand side of (6.33) *achieves its minimum at an equilibrium state.* Therefore, at such a state, its first differential vanishes, and its second differential is a positive-semidefinite form there, these differentials being obtained from functions so reduced that all expressed arguments are independent variables.

The first condition leads to explicit formulae for the stress and heating flux at an equilibrium state:

$$\begin{aligned} \bar{\mathbf{T}}(\theta^+, \mathbf{0}, \mathbf{F}^+, \mathbf{0}, \boldsymbol{\xi}^+) &= \rho^+\mathbf{F}^+[\partial_{\mathbf{F}}\bar{\psi}(\theta^+, \mathbf{F}^+, \boldsymbol{\xi}^+)^T \\ &\qquad - \bar{\boldsymbol{\alpha}}(\theta^+, \mathbf{F}^+, \boldsymbol{\xi}^+) * \partial_{\dot{\mathbf{F}}}\bar{\mathbf{g}}(\theta^+, \mathbf{0}, \mathbf{F}^+, \mathbf{0}, \boldsymbol{\xi}^+)^T], \\ \bar{\mathbf{h}}(\theta^+, \mathbf{0}, \mathbf{F}^+, \mathbf{0}, \boldsymbol{\xi}^+) &= \rho^+\bar{\boldsymbol{\alpha}}(\theta^+, \mathbf{F}^+, \boldsymbol{\xi}^+) * \partial_{\boldsymbol{\gamma}}\bar{\mathbf{g}}(\theta^+, \mathbf{0}, \mathbf{F}^+, \mathbf{0}, \boldsymbol{\xi}^+). \end{aligned} \tag{6.40}$$

By (6.39), we see from (6.40) that *at a strong equilibrium state, the heating flux vanishes and the free-energetic function* $\bar{\psi}$ *is a thermostatic potential for the stress.* At a false equilibrium state, however, neither conclusion follows in general. For a material in which the molar growth is independent of the temperature gradient, $\mathbf{h}$ vanishes even at a false equilibrium state; for a material in which the molar growth is independent of the velocity of deformation, $\bar{\psi}$ is a thermostatic potential for the stress even at such a state; more generally, it is still possible that the terms in question shall vanish at some false equilibrium states, but they will not do so at all such states. That is, *at a false equilibrium state the stress and heating flux need not have the values they would if that same state were a strong one.* Also

$$\begin{aligned} \bar{\boldsymbol{\alpha}}(\theta^+, \mathbf{F}^+, \boldsymbol{\xi}^+) * \partial_\theta\bar{\mathbf{g}}(\theta^+, \mathbf{0}, \mathbf{F}^+, \mathbf{0}, \boldsymbol{\xi}^+) &= 0, \\ \bar{\boldsymbol{\alpha}}(\theta^+, \mathbf{F}^+, \boldsymbol{\xi}^+) * \partial_{\mathbf{F}}\bar{\mathbf{g}}(\theta, \mathbf{0}, \mathbf{F}^+, \mathbf{0}, \boldsymbol{\xi}^+) &= 0, \\ [\partial_{\boldsymbol{\xi}}\bar{\mathbf{g}}(\theta^+, \mathbf{0}, \mathbf{F}^+, \mathbf{0}, \boldsymbol{\xi}^+)]^T\bar{\boldsymbol{\alpha}}(\theta^+, \mathbf{F}^+, \boldsymbol{\xi}^+) &= \mathbf{0}. \end{aligned} \tag{6.41}$$

These three conditions are trivially satisfied at a strong equilibrium state. More generally, they restrict the chemical affinity $\boldsymbol{\alpha}$ at a false equilibrium

† In the literature of chemistry, it seems, (6.38) by itself defines "false equilibrium", and the combined conditions (6.38) and (6.39) define "equilibrium". In the linear "irreversible thermodynamics" described in the next lecture, the two concepts are confused by linear "phenomenological relations" and talk about "forces" and "fluxes".

state. Namely, in terms of the component space $\mathcal{V}$, *the chemical affinity must be orthogonal to the values of* $\partial_\theta \bar{\mathbf{g}}$ *and* $\partial_\mathbf{F} \bar{\mathbf{g}}$ *and must lie in the kernel of* $(\partial_\xi \mathbf{g})^T$. These conditions restrict the value the chemical affinity $\boldsymbol{\alpha}$ may have at a false equilibrium state, but they need not annul it at all such states.

In most chemical theories, conditions of regularity so strong as to exclude false equilibrium are imposed at the outset. Suppose, for example, that the function $\mathbf{g}(\theta, \mathbf{0}, \mathbf{F}, \mathbf{0}, \boldsymbol{\xi})$ be smoothly invertible for $\boldsymbol{\xi}$ in a neighborhood of $(\mathbf{F}^+, \theta^+)$; that is, there exists a function $\bar{\boldsymbol{\xi}}(\theta, \mathbf{F})$ such that

$$\bar{\mathbf{g}}(\theta, \mathbf{0}, \mathbf{F}, \mathbf{0}, \bar{\boldsymbol{\xi}}(\theta, \mathbf{F})) = \mathbf{0} \tag{6.42}$$

in a neighborhood of $(\theta^+, \mathbf{F}^+)$. Invertibility such as this is thoughtlessly invoked in much of classical thermodynamics. If (6.42) holds, the kernel of $(\partial_\xi \bar{\mathbf{g}})^T$ at $(\theta^+, \mathbf{0}, \mathbf{F}^+, \mathbf{0}, \boldsymbol{\xi}^+)$ reduces to the zero vector alone, so the condition $(6.41)_3$ requires that (6.39) hold. Thus *an equilibrium state near which* $\bar{\mathbf{g}}$ *is smoothly invertible is never a false one.*

The conditions on the second differential at an equilibrium state are more complicated. There is no need to duplicate here the remainder of Bowen's memoir. It is enough to mention some of his conclusions. In particular, Bowen investigates states $(\theta', \mathbf{F}', \boldsymbol{\xi}')$ such that the chemical affinity vanishes:

$$\bar{\boldsymbol{\alpha}}(\theta', \mathbf{F}', \boldsymbol{\xi}') = \mathbf{0}, \tag{6.43}$$

and he asks if these be necessarily equilibrium states. He shows that in general they are not. That is, (6.43) does not imply (6.38). However, if both $\partial_\xi \bar{\boldsymbol{\alpha}}$ and $\partial_\xi \bar{\mathbf{g}}$ are regular linear transformations at the state $(\boldsymbol{\theta}, \mathbf{0}, \mathbf{F}, \mathbf{0}, \boldsymbol{\xi})$, then the two conditions (6.43) and (6.38) are equivalent there. Not only is every equilibrium state then a strong one, but also the vanishing of the chemical affinity becomes a necessary and sufficient condition for equilibrium.

Bowen considers also the stability of equilibrium states† and the restrictions imposed by frame-indifference and material symmetry.

So as to exclude various kinds of undesirable response, Bowen proposes for study also an *a priori* inequality for mixtures that includes as special cases both the convexity condition of Gibbs and the inequality of Coleman & Noll for the thermal equilibrium of simple materials‡,

† The stability theory of differential equations was first applied to chemistry, it seems, by J. Wei, "Axiomatic treatment of chemical reaction systems," *J. Chem. Phys.*, **36**:1578–1584 (1962), and was further developed by Coleman & Gurtin in their paper cited earlier in this lecture.

‡ *Cf.* Section 81 of *The Non-Linear Field Theories of Mechanics*, in Flügge's *Handbuch der Physik*, vol. III/3, Springer-Verlag, Berlin-Heidelberg-New York, 1965.

namely,

$$\psi(\theta, \mathbf{F}, \xi) - \psi(\theta', \mathbf{F}', \xi') - (\theta - \theta')\partial_\theta\psi(\theta, \mathbf{F}, \xi) \\ - \operatorname{tr}\{(\mathbf{F} - \mathbf{F}')\partial_{\mathbf{F}}\psi(\theta', \mathbf{F}', \xi')^{\mathrm{T}}\} \\ - (\xi - \xi') * \partial_\xi\psi(\theta', \mathbf{F}', \xi') > 0, \quad (6.44)$$

except in the trivial case when $\theta = \theta'$, $\mathbf{F} = \mathbf{F}'$, and $\xi = \xi'$, and provided $\mathbf{F}(\mathbf{F}')^{-1}$ be positive-definite and symmetric. He shows that at a state where the chemical affinity vanishes, this condition suffices to render the free energetic a minimum under variations of the extent of reaction alone and that the corresponding function giving the differences of free energetic is a Liapunov function. Thus for such materials definite theorems of stability of a state of strong equilibrium may be proved. In particular, if the inequality (6.44) is imposed on the free-energetic function ψ, *any state of strong equilibrium is stable with respect to reactions that take place at constant deformation and temperature.*

The decisive results gotten by extending the thermodynamic approach of Coleman & Noll to diffusion in non-reacting mixtures and to chemical reactions in non-diffusing mixtures, sketched in this lecture and the one before it, give fair promise for a rational science of chemical thermomechanics as a whole, with diffusion and reaction combined. Surely such a science is not far off now. For this series, however, I have finished my sketch of modern phenomenological thermodynamics, and the following four lectures will concern other theories of heat and temperature.

Pelagus quantos aperimus in usus!

Lecture 7

THE ONSAGER RELATIONS

To those who show with great brevity and clearness the fallacy of propositions commonly held true by the generality, it would be a fairly bearable injury to be repaid only by contempt instead of gratitude; but very unpleasant and grievous is a certain other sentiment which is sometimes aroused in some persons who claim to be at least the peers of anybody in the very same studies, persons who are seen to have maintained as true conclusions which another has then by a short and easy discourse unveiled and proclaimed false. I shall not name that sentiment envy, which usually turns into hate and anger against those who discovered the fallacies, but I will call it rather annoyance and yearning to be able to maintain hoary old errors rather than allow newly discovered truths to be received. This yearning sometimes induces them to write in contradiction of truths even they themselves admit in their hearts, just so as to keep down the fame of others in the opinion of the numerous and ill informed common folk.

Galileo

The classical theory of heat conduction, first proposed by Fourier and then generalized to crystalline bodies by Duhamel in 1828, rests on the constitutive assumption that the heating flux $\mathbf{h}$ is a linear function of the temperature gradient:

$$\mathbf{h} = \mathbf{K}(\theta, \mathbf{F}) \operatorname{grad} \theta, \tag{7.1}$$

where θ is the temperature and $\mathbf{F}$ is the gradient of the deformation from some fixed reference configuration. The tensor $\mathbf{K}$ is the *thermal conductivity*. A differential equation for the temperature field $\theta(\mathbf{x}, t)$ is obtained from the balance of energy:

$$\rho\dot{\varepsilon} = w + \operatorname{div} \mathbf{h} + \rho s, \tag{2.18}$$

where ε is the energetic, w is the stress power, and s is the body heating. This last, s, is to be regarded as a given function of position; in most special problems it is assumed to vanish. The classical theory refers, in the main, to a material obeying a caloric equation of state:

$$\varepsilon = \mathfrak{e}(\theta, \mathbf{F}, \mathbf{X}), \tag{7.2}$$

and it restricts itself to static deformation. Then

$$\begin{aligned} \dot{\varepsilon} &= c\partial_t\theta \qquad \text{where } c = c(\theta, \mathbf{F}, \mathbf{X}) \equiv \partial_\theta \mathfrak{e}, \\ w &= 0. \end{aligned} \tag{7.3}$$

Thus substitution of (7.1) into (2.18) when $s = 0$ yields

$$\begin{aligned} \rho c \partial_t \theta &= \operatorname{div}\,(\mathbf{K}\operatorname{grad}\theta), \\ &= \operatorname{tr}\,(\mathbf{K}_+ \operatorname{grad}^2\theta) + (\operatorname{div}\mathbf{K}^T)\operatorname{grad}\theta, \end{aligned} \tag{7.4}$$

where $\mathbf{K}_+$ is the symmetric part of $\mathbf{K}$:

$$\mathbf{K}_+ \equiv \tfrac{1}{2}(\mathbf{K} + \mathbf{K}^T). \tag{7.5}$$

A conductor of heat is said to be *homogeneous* if there exists a reference configuration such that $\mathbf{K}$ and $\mathfrak{c}$ do not depend on the reference position $\mathbf{X}$. In a homogeneous configuration of such a body, c and $\mathbf{K}$ reduce to functions of θ alone, and (7.4) becomes a differential equation for determination of θ:

$$\begin{aligned} \rho c \partial_t \theta &= \operatorname{tr}\,(\mathbf{K}_+ \operatorname{grad}^2\theta) + \operatorname{grad}\theta \cdot (\partial_\theta \mathbf{K}_+)\operatorname{grad}\theta, \\ c &= c(\theta), \qquad \mathbf{K}_+ = \mathbf{K}_+(\theta), \end{aligned} \tag{7.6}$$

since $\mathbf{F}$ and ρ now are given constants. In the common applications, c and $\mathbf{K}_+$ are taken as being independent of θ, so that the second term vanishes and (7.6) becomes a linear parabolic equation with constant coefficients, but this further simplification need not be made. The differential equation (7.6), as it stands, *is unaffected by* $\mathbf{K}_-$, *the skew part of the conductivity:*

$$\mathbf{K}_- \equiv \tfrac{1}{2}(\mathbf{K} - \mathbf{K}^T). \tag{7.7}$$

Therefore, if the boundary conditions are likewise independent of $\mathbf{K}_-$, there is no loss in generality if we simply *assume* that $\mathbf{K}$ is symmetric. In particular, *for a homogeneous rigid conductor of heat subject to prescribed temperatures upon its surface, no measurement of internal temperatures can determine whether or not* $\mathbf{K}$ *be symmetric.* Moreover, the classical *a priori* inequality for heat conduction

$$-\mathbf{h} \cdot \boldsymbol{\gamma} \geqq 0, \tag{2.46}$$

which asserts that heat does not flow from cold to hot, here takes the form

$$\mathbf{m} \cdot \mathbf{K}_+ \mathbf{m} \geqq 0 \qquad \text{for all } \mathbf{m}. \tag{7.8}$$

That is, the symmetric part of $\mathbf{K}$ is positive-semidefinite. Again, only the symmetric part $\mathbf{K}_+$ enters the statement.

Most of the foregoing observations are taken from a great note by STOKES, published in 1851. STOKES went on to remark that for many types of crystalline symmetry, the skew part $\mathbf{K}_-$ of the conductivity can be proved to vanish. Indeed, it is possible to show that for a solid referred to an undistorted state, an orthogonal tensor $\mathbf{Q}$ belongs to the symmetry group of the material if and only if it commutes with $\mathbf{K}$:

$$\mathbf{QK} = \mathbf{KQ}. \tag{7.9}$$

When a class of crystals is laid down, the most general symmetries of K's possible for it may be determined from this equation. As far as heat conduction alone is concerned, there are thirteen types of crystals. For seven of these, **K** must be symmetric. These seven include all the simpler kinds: isotropic materials, crystals with two planes of symmetry, and even some with no planes of symmetry, as STOKES observed. Specifically, of the thirty-two optical crystal classes, nineteen are such as to force **K** to be symmetric. Moreover, in only two classes, namely, those of the triclinic system, may $\mathbf{K}_-$ be arbitrary; in all the others in which $\mathbf{K}_-$ need not vanish entirely, at least one of its off-diagonal components with respect to a preferred set of axes must vanish. (An exhaustive modern treatment by C.-C. WANG, based on results of COLEMAN & NOLL, is given as an appendix to this lecture.) Thus, as STOKES wrote of the non-zero components of $\mathbf{K}_-$, "it is only among crystals which possess a peculiar sort of asymmetry, that we should expect to find traces of their existence."

When it is so difficult to find an experiment whose outcome is affected by the nature of $\mathbf{K}_-$, it would seem that whether or not $\mathbf{K}_-$ vanishes in the few cases when it does not have to by reasons of crystalline symmetry, might be of small interest, but the contrary is true. STOKES offered the conjecture that *in all bodies*

$$\mathbf{K}_- = \mathbf{0}. \tag{7.10}$$

In favor of it, he adduced the following remark. By (7.4), the steady distributions of temperature in a body with constant conductivity **K** satisfy

$$\operatorname{tr}(\mathbf{K}_+ \operatorname{grad}^2 \theta) = 0. \tag{7.11}$$

A particular solution, considered to represent a "source of heat at the origin", is

$$\theta = \frac{\text{const.}}{\sqrt{\mathbf{p} \cdot \mathbf{K}_+^{-1}\mathbf{p}}} \tag{7.12}$$

where **p** is the position vector with respect to the origin. For this solution, the isotherms are the concentric ellipsoids

$$\mathbf{p} \cdot \mathbf{K}_+^{-1}\mathbf{p} = \text{const.}; \tag{7.13}$$

the temperature gradient is parallel to $\mathbf{K}_+^{-1}\mathbf{p}$, and the heating flux is

$$\begin{aligned} \mathbf{h} &= (\mathbf{K}_+ + \mathbf{K}_-) \operatorname{grad} \theta, \\ &\propto \mathbf{p} + \mathbf{K}_-\mathbf{K}_+^{-1}\mathbf{p}. \end{aligned} \tag{7.14}$$

If $\mathbf{K}_- = \mathbf{0}$, the lines of heating flux are radial; if $\mathbf{K}_- \neq \mathbf{0}$, they are not, and the energy, if pictured as a fluid, flows along curved streamlines. STOKES wrote, "This *rotatory* sort of motion of heat, produced by the mere diffusion from the source outwards, certainly seems very strange, and leads us to think" that $\mathbf{K}_- = \mathbf{0}$ in all cases.

Experiments were undertaken to see whether a material with non-symmetric $\mathbf{K}$ could be found. Necessarily, as just shown, they had to concern a suitably unsymmetric crystal and a boundary condition other than that of prescribed temperature. At a calorically insulated boundary, $\mathbf{n} \cdot \mathbf{h} = 0$, so DUHAMEL's law (7.1) gives

$$\mathbf{n} \cdot (\mathbf{K}_+ + \mathbf{K}_-) \operatorname{grad} \theta = 0, \tag{7.15}$$

where $\mathbf{n}$ is the unit normal. In principle, the experiment goes as follows. The experimenter first determines $\mathbf{K}_+$ for the crystal by measurement of temperatures. Then he insulates the boundary and determines the isotherms near it, so that the direction of grad θ is known. Thus he can see whether or not

$$\mathbf{n} \cdot \mathbf{K}_+ \operatorname{grad} \theta = 0 \tag{7.16}$$

there. If so, $\mathbf{n} \cdot \mathbf{K}_- \operatorname{grad} \theta = 0$ also, but in order to conclude that $\mathbf{K}_- = \mathbf{0}$, he has to be sure that grad θ has non-zero components in at least two tangential directions for one boundary surface and in at least one tangential direction for an orthogonal boundary surface.

Experiments to test the symmetry of $\mathbf{K}$ were conceived by SORET, P. CURIE, and VOIGT†; SORET made measurements in gypsum, erythrite, dolomite, and apatite; VOIGT, in the last two. They did not find evidence of dissymmetry. Since then, it has been regarded as an experimental fact that $\mathbf{K}_- = \mathbf{0}$ in all materials. I should overstep the bounds of my competence were I to give voice to any uncertainty that these experiments measure what they were claimed to, but even theorists are allowed to count, and hence I may mention that so far as I know, these experiments have not been checked by repetition, nor have tests been reported on any specimens from the eleven further types of crystals in which $\mathbf{K}_-$ need not vanish by symmetry, or in transversely isotropic materials‡.

† CH. SORET, "De la conductibilité calorifique dans les cristaux," *J. de Physique*, (3) **2**:241–259 (1893) = pp. 482–504 of his *Éléments de cristallographie physique*, Gauthier-Villars, Paris, 1893, with an additional note.

P. CURIE, "À propos des éléments de cristallographie physique de M. Ch. Soret," *Arch. Soc. Phys. Hist. Nat. Genève*, **29**:237–354 (1893).

CH. SORET, "Sur l'étude expérimentale des coefficients rotationnels de conductibilité thermique," *ibid.*, 355–357.

Procès-verbaux, *ibid.*, 322–323 and **32**:631–633 (1894).

W. VOIGT, "Fragen der Kristallphysik, I," *Nachr. Ges. Wiss. Göttingen*, 1903, No. 3, 87–89.

‡ GURTIN has kindly let me see a manuscript in which he obtains a phenomenological interpretation of the symmetry of $\mathbf{K}$. He formulates a concept of thermal stability and shows that in terms of it, a rigid heat conductor is thermally stable if and

The next theory of a dissipative material to be proposed was the Navier-Poisson-Stokes theory of viscous fluids (1821–1845). Here the stress $\mathbf{T}$ is supposed to be a linear function of the velocity gradient:

$$\mathbf{T} = \mathsf{V}\, \mathrm{grad}\, \dot{\mathbf{x}}, \tag{7.17}$$

where V is a fourth-order tensor called the *viscosity operator*. While all the early authors laid down an assumption that the fluid be isotropic, we now know that no such assumption is needed. If we assume merely that

$$\mathsf{V} = \mathsf{V}(\dot{\mathbf{x}}, \mathbf{x}, t), \tag{7.18}$$

then from the principle of material frame-indifference, which asserts that the response of the material is independent of the observer, it can be proved† that V is a constant isotropic tensor and hence that (7.17) reduces to the classical form

$$\mathbf{T} = -p\mathbf{1} + \mathbf{V}, \tag{2.31}$$

---

only if $\mathbf{K}_- = \mathbf{0}$, $\mathbf{K}_+$ is positive-semidefinite, and the specific heat c is non-negative. Thus the symmetry of $\mathbf{K}$ joins the usual inequalities of the theory of heat conduction as a condition with a definite gross meaning. However, GURTIN's concept of stability does not seem to be related to any general principle of physics but rather is an additional condition. It may point the way toward some future success in giving the "Onsager relations" at least a meaningful, self-consistent *statement* for field phenomena.

Note added in March, 1969. The work just described has been published in more general form by W. A. DAY & M. E. GURTIN, "On the symmetries of the conductivity tensor and other restrictions in the nonlinear theory of heat conduction," *Arch. Rational Mech. Anal.*, **33**:26–32 (1969).

† W. NOLL, "On the continuity of the solid and fluid states," *J. Rational Mech. Anal.*, **4**:3–81 (1955) = pp. 65–143 of *Rational Mechanics of Materials*, Gordon & Breach, New York, 1965. See Section 13B.

Some of the literature purports to deal with "anisotropic fluids" obeying (7.17) with a V which is not isotropic; *e.g.* (following blindly the formal procedures of linear elasticity), in Section I.3 of the survey article by H. FALKENHAGEN, "Klassische Hydrodynamik," WIEN & HARMS' *Handbuch der Experimentalphysik*, vol. 4, part 1, Akademische Verlagsgesellschaft, Leipzig, 1931.

Theories of anisotropic materials of this kind are simply wrong, since they allow a rigid motion of the body as a whole to change the way in which its parts react on one another. If, for example,

$$\mathbf{T} = \mathsf{V}\mathbf{D},$$

then a rotation $\mathbf{Q}$ transforms $\mathbf{D}$ into $\mathbf{QDQ}^T$ and should transform $\mathbf{T}$ into $\mathbf{QTQ}^T$, but unless the fourth-order tensor V is isotropic, it does not.

Of course, certain real materials show effects to which the vague language of physics refers as being both "fluid" and "anisotropic". To describe and predict such effects, mechanically sound theories have been proposed, notably the "anisotropic fluids" of ERICKSEN, the "simple subfluids" of WANG, and the "simple liquid crystals" of COLEMAN, but these theories rest on constitutive equations not of the type (7.17) and hence not open to the objection just raised.

where

$$\mathbf{V} = \lambda(\mathrm{tr}\,\mathbf{D})\mathbf{1} + 2\mu\mathbf{D},$$

and $2\mathbf{D} = \mathrm{grad}\,\dot{\mathbf{x}} + (\mathrm{grad}\,\dot{\mathbf{x}})^T$. Therefore, as a consequence of general physical principles, $\mathbf{V}$ is not only a symmetric tensor but also one of a very special kind.

Almost as old as the theories of heat conduction and viscosity is the third theory of a dissipative material, namely, FICK's theory of diffusion (1855). While originally it was too primitive to reveal any ideas of principle, MAXWELL soon gave it a rational basis in his kinetic theory of gas mixtures, and STEFAN in a major paper† of 1871 cleared away the specifically kinetic details so as to achieve an inclusive phenomenological theory. We may describe his proposal in modern terms as a constitutive equation for the momentum growth $\underset{\mathfrak{a}}{\overset{+}{\mathbf{m}}}$ of the $\mathfrak{a}^{\text{th}}$ constituent of a mixture (*cf.* $(5.10)_1$):

$$\rho\underset{\mathfrak{a}}{\overset{+}{\mathbf{m}}} = -\sum_{\mathfrak{b}=1}^{\mathfrak{k}} F_{\mathfrak{a}\mathfrak{b}}(\underset{1}{\rho}, \underset{2}{\rho}, \ldots, \underset{\mathfrak{k}}{\rho})(\underset{\mathfrak{a}}{\dot{\mathbf{x}}} - \underset{\mathfrak{b}}{\dot{\mathbf{x}}}), \tag{7.19}$$

where $\underset{\mathfrak{c}}{\rho}$ and $\underset{\mathfrak{c}}{\dot{\mathbf{x}}}$ are the density and the velocity of the constituent $\mathfrak{c}$. The *diffusive drag coefficient* $F_{\mathfrak{a}\mathfrak{b}}$ is the rate of loss of momentum of constituent $\mathfrak{a}$ per unit velocity of its motion relative to constituent $\mathfrak{b}$ and per unit volume. If $\underset{\mathfrak{a}}{\overset{+}{\mathbf{m}}}$ is a known linear function of the $\underset{\mathfrak{b}}{\dot{\mathbf{x}}}$, the coefficients $F_{\mathfrak{a}\mathfrak{b}}$ in (7.19) are not uniquely determined, but if the $\overset{*}{F}_{\mathfrak{a}\mathfrak{b}}$ constitute a second possible set, then there exists a function $F(\underset{1}{\rho}, \underset{2}{\rho}, \ldots, \underset{\mathfrak{k}}{\rho})$ such that

$$F_{\mathfrak{a}\mathfrak{b}} - \overset{*}{F}_{\mathfrak{a}\mathfrak{b}} = F\delta_{\mathfrak{a}\mathfrak{b}}. \tag{7.20}$$

Any statement about the $F_{\mathfrak{a}\mathfrak{b}}$, therefore, must be invariant under a transformation of this kind. For a binary mixture, MAXWELL appealed to the principle of action and reaction to infer that

$$F_{12} = F_{21}. \tag{7.21}$$

If the mixture has more than two constituents, diffusion need not be the effect of interaction of pairs alone, and even the most credulous allowance to "Newton's third law" cannot yield here a general "reciprocity relation":

$$F_{\mathfrak{a}\mathfrak{b}} = F_{\mathfrak{b}\mathfrak{a}}. \tag{7.22}$$

By appeal to the principle of linear momentum, STEFAN deduced (7.21) for a binary mixture. Without giving any further reason, he then asserted

† J. STEFAN, "Über das Gleichgewicht und die Bewegung, insbesondere die Diffusion von Gasmengen," *Wiener Sitzungsberichte*, **63**:63–124 (1871).

(7.22). Accordingly, (7.22) deserves to be named "STEFAN's reciprocal relations".

While, as has often been the case with great masters, MAXWELL and STEFAN drew back just short of error, more recent authors have not always been so careful or so lucky: For mixtures of three or more constituents, some have appealed to one or the other mechanical principle to "prove" STEFAN's relations. In fact, the principle of linear momentum, as we have seen in the preceding lecture, requires only that

$$\sum_{\mathfrak{a}} \overset{+}{\mathfrak{m}} = \mathbf{0}. \tag{5.12$_2$}$$

For the constitutive equations (7.19), this condition is equivalent to†

$$\sum_{\mathfrak{a}=1}^{\mathfrak{k}} (\digamma_{\mathfrak{ab}} - \digamma_{\mathfrak{ba}}) = 0, \qquad \mathfrak{b} = 1, 2, \ldots, \mathfrak{k}. \tag{7.23}$$

For a binary mixture, this condition is equivalent to (7.21). For a mixture of three or more constituents, it is not so strong as (7.22). STEFAN's relations imply the balance of linear momentum, but it does not imply them if $\mathfrak{k} > 2$.

As far as I can learn, the first experimental tests of STEFAN's relations were made by DUNLOP & GOSTING‡ in 1955. As I read the results reported by them and others for eleven different solutions in water, they indicate that for eight systems STEFAN's relations hold, that for one they are doubtful, and that for two they clearly do not hold. Nevertheless D. G. MILLER in his authoritative survey paper§ finds "the agreement . . . remarkably good" and concludes that STEFAN's relations are confirmed by experiment.

Other "reciprocal relations" in irreversible processes were investigated in the last century and this, most of them involving electromagnetic effects. In the rest of this lecture I shall discuss only the general theory and its bearing on the three phenomena already mentioned: conduction of heat, viscosity, and diffusion.

The classical theories of these three phenomena may be set upon a

† Apparently TOUPIN & I were the first to publish this observation. See Section 295 of *The Classical Field Theories*, in FLÜGGE's *Handbuch der Physik*, vol. III/1, Springer-Verlag, 1960; *cf.* also my papers, "Una teoria meccanica della diffusione," *Celebrazioni Archimedee del secolo XX* (Siracusa, 1961), **3**:161–168 (1962), and "Mechanical basis of diffusion," *J. Chem. Phys.*, **37**:2336–2344 (1962).

‡ P. J. DUNLOP & L. J. GOSTING, "Interacting flows in liquid diffusion: Expressions for the solute concentration curves in free diffusion, and their use in interpreting Gouy diffusiometer data for aqueous three-component systems," *J. Am. Chem. Soc.*, **77**:5238–5249 (1955).

§ D. G. MILLER, "Thermodynamics of irreversible processes, the experimental verification of the Onsager reciprocal relations," *Chem. Rev.*, **60**:15–37 (1960).

thermodynamic framework. If we introduce the "entropy production" or *caloric growth* Δ:

$$\Delta \equiv \rho\delta - \mathbf{h} \cdot \boldsymbol{\gamma}, \tag{7.24}$$

where δ is the internal dissipation as defined by (2.25), and where $\boldsymbol{\gamma} \equiv (1/\vartheta) \text{ grad } \vartheta = (-1/\theta) \text{ grad } \theta$, then†

$$\Delta = \begin{cases} -\boldsymbol{\gamma} \cdot \mathbf{h} & \text{for heat conduction alone,} \\ \text{tr}\,(\mathbf{VD}) & \text{for viscosity alone,} \\ -\rho\Sigma \overset{+}{\underset{a}{\mathbf{m}}} \cdot \underset{a}{\dot{\mathbf{x}}} & \text{for diffusion alone.} \end{cases} \tag{7.25}$$

That is,

$$\Delta = \boldsymbol{\xi} \cdot \boldsymbol{\iota}, \tag{7.26}$$

where $\boldsymbol{\xi}$ and $\boldsymbol{\iota}$ are vectors of 3, 6, and $3\mathfrak{k} - 3$ dimensions,‡ respectively, for the three different theories. The constitutive equations are of the form

$$\boldsymbol{\iota} = \mathbf{\Lambda}\boldsymbol{\xi}, \tag{7.27}$$

where $\mathbf{\Lambda}$ is a constant tensor. Hence Δ is given by a quadratic form in $\boldsymbol{\xi}$:

$$\Delta = \Delta(\boldsymbol{\xi}) = \boldsymbol{\xi} \cdot \mathbf{\Lambda}\boldsymbol{\xi} = \boldsymbol{\xi} \cdot \mathbf{\Lambda}_{+}\boldsymbol{\xi}. \tag{7.28}$$

The Clausius-Duhem inequality (2.47) is satisfied identically, therefore, if and only if $\mathbf{\Lambda}_{+}$, the symmetric part of $\mathbf{\Lambda}$, is positive-definite. Thus $\mathbf{\Lambda}_{-}$, the skew part of $\mathbf{\Lambda}$, contributes nothing to the caloric growth. Also

$$\mathbf{\Lambda}_{-} = \mathbf{0} \begin{cases} \text{in heat conduction for the simpler crystalline solids (in virtue of material symmetry),} \\ \text{in viscosity always (in virtue of frame-indifference),} \\ \text{in diffusion of a binary mixture (in virtue of the balance of momentum)} \end{cases} \tag{7.29}$$

and

$$\mathbf{\Lambda}_{-} = \mathbf{0} \begin{cases} \text{in heat conduction for all materials if STOKES's conjecture holds,} \\ \text{in diffusion of any mixture if STEFAN's relations hold.} \end{cases} \tag{7.30}$$

† The result for diffusion is not proved in Lecture 5 but is easily seen from (5.11) and (5.12).

‡ The relation $\Sigma \overset{+}{\underset{a}{\mathbf{m}}} = 0$ is used to eliminate one of the $\overset{+}{\underset{a}{\mathbf{m}}}$. For simplicity I presume in all further remarks about diffusion that the variables have been reduced in this way so as to form a linearly independent set of vectors.

In a paper† published in 1931 ONSAGER attempted to correlate these observations by constructing a general theory of linear constitutive equations of this kind. A colossal literature has grown from it. ONSAGER claimed to prove "a general class of reciprocal relations in irreversible processes from the *assumption of microscopic reversibility.*" For the class of processes he considered, his result is

$$\mathbf{\Lambda}_{-} = \mathbf{0} \tag{7.31}$$

without exception. This statement, after later modification to take account of time reversal and magnetic fields, is called "ONSAGER's reciprocal relations". The particular phenomenon ONSAGER chose for detailed analysis was heat conduction, and he concluded, unconditionally, that STOKES's conjecture (7.10) is true:

$$\mathbf{K} = \mathbf{K}^{\mathrm{T}}. \tag{7.32}$$

ONSAGER warned the reader that his more inclusive inference (7.31) could not be expected to hold for "systems where external magnetic fields are acting‡, and systems whose motion is described relatively to a rotating frame of coordinates . . . ." This last statement shows that he did not adopt the principle of material frame-indifference§: He believed that the response of a material is frame-bound, different for observers in different frames. This strange notion seems to have diffused through all the literature that has grown from ONSAGER's paper.

ONSAGER's reasoning and results were criticized by CASIMIR¶ in 1945. Like ONSAGER before him, CASIMIR chose the conduction of heat in crystals as his main application. His result is weaker than ONSAGER's,

† L. ONSAGER, "Reciprocal relations in irreversible processes," *Phys. Rev.*, (2) **37**:405–426 (1931); **38**:2265–2279 (1932).

‡ In the second of the papers cited, ONSAGER considered the case when **K** is allowed to depend on the magnetic field **B**. He inferred that (7.32) should then be replaced by

$$\mathbf{K}(-\mathbf{B}) = \mathbf{K}(\mathbf{B})^{\mathrm{T}}.$$

§ While the name and the general statement of the principle of material frame-indifference are recent, major special cases were proposed and applied by POISSON (1829), CAUCHY (1829), ZAREMBA (1903), and JAUMANN (1906). For discussion and history, see Section 19A of *The Non-Linear Field Theories of Mechanics,* in FLÜGGE's *Handbuch der Physik,* vol. III/3, Springer-Verlag, 1965.

¶ H. B. G. CASIMIR, "On Onsager's principle of microscopic reversibility," *Rev. Mod. Phys.*, **17**:343–350 (1945). CASIMIR wrote, ". . . ONSAGER has tried to show . . . [*etc.*]. But although we do not doubt the essential truth of ONSAGER's ideas, his application of these ideas to particular cases is in our opinion not always entirely satisfactory."

Cf. also L. ONSAGER & S. MACHLUP, "Fluctuations and irreversible processes," *Phys. Rev.*, (2) **91**:1505–1515 (1953).

namely,

$$\operatorname{div} \mathbf{K}_- = \mathbf{0} \tag{7.33}$$

rather than $\mathbf{K}_- = \mathbf{0}$. In a homogeneous crystal, $\mathbf{K}$ is a function of θ alone, so CASIMIR's reciprocal relation (7.33) reduces to

$$(\partial_\theta \mathbf{K}_-) \operatorname{grad} \theta = \mathbf{0} \tag{7.34}$$

for all temperature fields. Hence his result requires that $\partial_\theta \mathbf{K}_- = \mathbf{0}$; that is, $\mathbf{K}_-$ is a constant†, but it need not be zero. Thus if we accept CASIMIR's analysis, we must conclude that heat conduction in a body of uniform conductivity joins viscosity and binary diffusion as a theory in which the reciprocal relations say nothing that is not known already *a fortiori*. In explanation of the difference between his result and ONSAGER's, CASIMIR wrote "our result is, that an antisymmetric component of $\mathbf{K}$ has no observable physical consequence whatever and it is permissible to put it equal to zero. It must be zero if we agree that $\mathbf{K} = \mathbf{0}$ in a vacuum and that $\mathbf{K}$ for a given substance does not depend on the shape of the sample. Though it would of course be very foolish not to accept this convention, it is interesting to note that ONSAGER's relation by itself does not compel us to do so." While I confess myself unable to see what the vacuum and the shape of the sample have to do with the tensor field $\mathbf{K}(\theta, \mathbf{F})$ in the constitutive equation for a rigid linear conductor of heat, I do recognize in CASIMIR's claim that only the divergence of $\mathbf{K}$ has "physical meaning" an incomplete rediscovery of the argument of STOKES, set forth at the beginning of this lecture. CASIMIR did not remark that if his claim were true, not only would the great experimenters SORET and VOIGT stand condemned of having claimed to measure something that has no "physical meaning", but also one of the main alleged experimental tests of the Onsager relations would reduce to vacuity.

Ever since that day the Onsagerists‡, as we call the adherents of linear phenomenological relations whose coefficients are subject to univer-

† This fact seems not to have been noticed in the literature, which resorts to evasions at (7.33) itself, not attempting to reduce it.

‡ This term is meant to be no more than descriptive and delimiting. While the Onsagerists describe their field as "the thermodynamics of irreversible processes", or, in the letter jargon of modern bureaucracy, as "TIP", to accept such a title would be to allow one particular twist of thought to arrogate an entire, major field of science, much as in politics the first man to grab the label "democratic" or "civil rights" for his own slant has half won the argument before it starts.

While some tender souls have been offended by the term "Onsagerist", there is good precedent for it. There have been Cartesians and Newtonians, and those who criticized the schools so named did not always mean to attack DESCARTES or NEWTON, nor were the high honors showered upon these savants themselves any guarantee of immunity for those who volunteered to huddle under their banners.

sal rules of symmetry, have had to face a trilemma on heat conduction in interpreting the words of the principal founders of their subject. First, they can agree with CASIMIR entirely and thus lose one of the main Onsager relations and the experimental support for it. Second, they can accept CASIMIR's reciprocal relation div $\mathbf{K}_- = \mathbf{0}$ as the correct one, but reject his claim that the value of $\mathbf{K}_-$ is of no physical significance; then they must admit that their theory is not strong enough to conclude the results that follow from material symmetry for most crystals, from the experiments of SORET and VOIGT for some others. Third, they can forget about CASIMIR and stay with ONSAGER. This last alternative was chosen by D. MILLER in his survey of experimental results; also, as was necessary in order to rehabilitate VOIGT and SORET, he refuted CASIMIR's claim that only div **h** has physical meaning, not **h** itself. The authors of the generally accepted master treatise of Onsagerism, DE GROOT & MAZUR†, give the reader a chance to swear by them for two of the three: On page 73 they present their own analysis leading to CASIMIR's result and assert that only div **h** has "physical meaning", but on page 237 they praise "the oldest and very accurate experiments" of SORET and VOIGT, from which they conclude that $\mathbf{K}_- = \mathbf{0}$ "was experimentally found to be valid."

ONSAGER's and CASIMIR's claims that their results follow from the "principle of microscopic reversibility" have been accepted with little question, especially by those who would never dream of trying to go through the mathematics and check the "proofs". In fact, it requires no great mathematician to see that the reversibility theorem and POINCARÉ's recurrence theorem make irreversible behavior impossible for dynamical systems in the classical sense. Something must be added to the dynamics of conservative systems, something not consistent with it, in order to get irreversibility at all. Everyone competent in mechanics has known this for a long time, but while in earlier work it was fashionable to shut one's eyes to the paradoxes and slip in the modifications tacitly, the difficulties of modern statistical mechanics have impelled students to look for and render explicit the alterations, often called "approximations", that they introduce. It would not be tactful for me to admit that I am not convinced by the mathematics‡ of ONSAGER and CASIMIR. I content myself with recalling that, details aside, both were fairly clear in stating some assumptions *beyond* "microscopic reversibility". These are§ that the distribution

† S. R. DE GROOT & P. MAZUR, *Non-equilibrium Thermodynamics*, North Holland, Amsterdam, 1962.

‡ In mathematics, if a proof is not found convincing by other mathematicians, the truth of the asserted theorem is subjected to doubt. In Onsagerist thermodynamics, anyone who doubts an alleged mathematical proof is contemptuously dismissed as being deficient in physical intuition.

§ *Cf.* the presentation in DE GROOT & MAZUR's treatise, Chapter VII.

function is Gaussian, that the logarithm of the density is proportional to the entropy in all cases (not only in equilibrium), that all macroscopic observables obey ordinary differential equations with constant coefficients as is "verified experimentally over a wide range of conditions"†, and, most important of all, that the expectations to be identified with macroscopic observables obey those same differential equations. Thus not even the *form* of the constitutive equations is derived from the molecular theory; rather, the molecular theory, so-called, is *forced into agreement* with preconceived phenomenological ideas. This is true, in particular, of CASIMIR's treatment of heat conduction. Even if we grant all the dubious mathematical steps, what comes out there is that *if* the molecular theory is consistent with the FOURIER-DUHAMEL theory, then div $\mathbf{K}_- = \mathbf{0}$, but we are left wondering whether any such consistency be possible at all.

My purpose here is not to question the special hypotheses of ONSAGER and CASIMIR but to emphasize the fact that they are made. The only one not likely to be questioned is microscopic reversibility; perhaps this is the reason that only this one, generally, is mentioned by the Onsagerists. We may well ask how important it is among the rest. Perhaps it is of no importance whatever, since what seems to be the most general and elegant proof, that of UHLHORN‡, does not mention it at all. Indeed, it seems that the Onsager-Casimir theorem belongs not to dynamics but rather to the theory of probability. Some assumptions of a purely stochastic kind are made plausible as statements of how things are, rather than derived as consequences of the laws of physics. The postulates of ONSAGER and CASIMIR may be good ones, but of course such an array of assumptions makes it impossible to regard their claims as having been *proved* from statistical mechanics in the sense that the theorems of MAXWELL, BOLTZMANN, GIBBS, and KHINCHIN are proved theorems.

Since, then, the assertions of ONSAGER and CASIMIR are not proved theorems of mathematical physics, they must be regarded as new principles, perhaps suggestive and useful, but, being postulates, necessarily open to experimental confirmation or disproof. In order to test them, we must first be able to state them in terms accessible, at least in principle, to experiment. One§ of the two standard formulations is as follows¶. Suppose the "entropy" or calory $\mathfrak{H}$ be a quadratic function of real m-dimensional

† P. 100 of DE GROOT & MAZUR's treatise. Note also their footnote xx, which in effect dismisses all the partial differential equations of mathematical physics by a reference which, when traced, leads back to CASIMIR's treatment of heat conduction.

‡ U. UHLHORN, "On the foundations of the linear theory of irreversible processes. I, II," *Arkiv Fysik*, **17**:257–272, 273–314 (1960).

§ The other, variational, is equivalent to it.

¶ The neat and inclusive argument here is due to U. UHLHORN, Section 2 of his paper just cited, condensing and generalizing many earlier arrangements.

vectors $\boldsymbol{\alpha}$; for example, the state $(\theta, \boldsymbol{\Upsilon})$ as described and employed in Lecture 1. The value $\boldsymbol{\alpha} = \mathbf{0}$ is called "the equilibrium state", and it is assumed to maximize $\mathfrak{H}(\boldsymbol{\alpha})$. Thus

$$\mathfrak{H}(\boldsymbol{\alpha}) - \mathfrak{H}(\mathbf{0}) = -\tfrac{1}{2}\boldsymbol{\alpha} \cdot \mathbf{S}\boldsymbol{\alpha}, \tag{7.35}$$

where $\mathbf{S}$ is a given, constant, positive-definite and symmetric tensor. In a thermokinetic process,

$$\boldsymbol{\alpha} = \boldsymbol{\alpha}(t), \tag{7.36}$$

and accordingly $\mathfrak{H} = \mathfrak{H}(\boldsymbol{\alpha}(t))$. The functions $\boldsymbol{\alpha}(t)$ are assumed to have a definite behavior under time reversal, which is represented in the theory by a real involutory linear transformation $\mathbf{T}$ over the space of vectors $\boldsymbol{\alpha}$:

$$\mathbf{T}^2 = \mathbf{1}. \tag{7.37}$$

(If we choose a basis, as we may, such that the matrix of $\mathbf{T}$ is a diagonal matrix whose entries T_{mm} are $+1$ or -1, the basis vector $\mathbf{e}_{\mathrm{m}}$ is called an "even variable" if $T_{\mathrm{mm}} = +1$ and an "odd variable" if $T_{\mathrm{mm}} = -1$.) The calory is assumed to be invariant under time reversal:

$$\mathfrak{H}(\mathbf{T}\boldsymbol{\alpha}) = \mathfrak{H}(\boldsymbol{\alpha}). \tag{7.38}$$

Hence (7.35) yields

$$\mathbf{T}^{\mathrm{T}}\mathbf{S}\mathbf{T} = \mathbf{S}. \tag{7.39}$$

The *thermodynamic force* $\boldsymbol{\xi}$ corresponding to the state $\boldsymbol{\alpha}$ is defined by

$$\boldsymbol{\xi} = \boldsymbol{\xi}(\boldsymbol{\alpha}) \equiv \partial_{\boldsymbol{\alpha}}\mathfrak{H}(\boldsymbol{\alpha}) = -\mathbf{S}\boldsymbol{\alpha}. \tag{7.40}$$

Hence

$$\boldsymbol{\xi}(\mathbf{0}) = \mathbf{0}, \qquad \boldsymbol{\alpha} = -\mathbf{S}^{-1}\boldsymbol{\xi}. \tag{7.41}$$

The *thermodynamic flux* $\boldsymbol{\iota}$ corresponding to the state $\boldsymbol{\alpha}$ is its time-rate of change:

$$\boldsymbol{\iota} \equiv \dot{\boldsymbol{\alpha}}. \tag{7.42}$$

Hence

$$\dot{\mathfrak{H}} = (\partial_{\boldsymbol{\alpha}}\mathfrak{H}) \cdot \dot{\boldsymbol{\alpha}} = \boldsymbol{\xi} \cdot \boldsymbol{\iota}. \tag{7.43}$$

That is, the rate of increase of calory is the inner product of the force and the flux. The class of constitutive equations considered is linear, of the same kind as we have studied in the theories of heat conduction, viscosity, and diffusion:

$$\boldsymbol{\iota} = \boldsymbol{\Lambda}\boldsymbol{\xi}, \tag{7.27}$$

where the "phenomenological coefficient" $\boldsymbol{\Lambda}$ is non-singular. Thus the flux is an invertible linear function of the force. The asserted "Onsager-

Casimir reciprocity relation" is

$$\mathbf{T\Lambda} = (\mathbf{T\Lambda})^{\mathrm{T}}: \tag{7.44}$$

The time-reversed phenomenological coefficient is symmetric. (If $\mathbf{T} = \mathbf{1}$, all the $\mathbf{e}_m$ are "even", and $\mathbf{\Lambda}$ is symmetric; this was the case treated in the early papers†.)

(I have left magnetic fields out of account, but I will say something about them at the end of this lecture.)

In this purely phenomenological theory, the symmetry relation is asserted, not motivated, but it loses none of its finality. If we can find a *single* system described by the postulates (7.35), (7.39), and (7.27), with $\boldsymbol{\xi}$ and $\boldsymbol{\iota}$ defined by (7.40) and (7.42), such that $\mathbf{T\Lambda}$ is not symmetric, then the Onsager-Casimir theory is disproved by experiments, for if (7.44) is not *general*, then there is nothing left but the statement that in linear constitutive relations, some tensor coefficients are symmetric and some are not. Thus, for example, if the theory as formulated applies to diffusion, and if STEFAN's relations fail for a *single* ternary mixture, then the whole Onsager-Casimir theory falls. It is not enough to conclude, as MILLER does in his survey, that in all but one solution the data confirm the assertion: *All* cases must agree.

However, the Onsagerists are too clever to offer us so easy an experimental test. Of the two "ifs" just stated, the former is the more difficult. In fact, the theory as just presented does not apply to *any* phenomenon described by fields, and in particular it does not apply to heat conduction, viscosity, or diffusion.

The essential definition is (7.42), according to which *the flux* $\boldsymbol{\iota}$ *is the time derivative of the state* $\boldsymbol{\alpha}$, where $\boldsymbol{\alpha}$ *is the argument of the caloric equation of state*, $\mathfrak{H} = \mathfrak{H}(\boldsymbol{\alpha})$. For conduction of heat in a rigid body, for example,

$$\begin{aligned} \dot{\mathfrak{H}} &= \int \dot{\eta}\, dM = \int \Delta\, dv, \\ \Delta &= -\mathbf{h}\cdot\boldsymbol{\gamma} = \frac{1}{\theta}\mathbf{h}\cdot \operatorname{grad}\theta, \\ \eta &= \eta(\mathbf{F}, \theta). \end{aligned} \tag{7.45}$$

If the constitutive relation is

$$\mathbf{h} = \mathbf{K}(\theta, \mathbf{F}) \operatorname{grad}\theta, \tag{7.1}$$

† These papers also made much of the fact that if $\mathbf{\Lambda}$ is symmetric, $\Delta(\boldsymbol{\xi})$ is a potential or "dissipation function" for the flux:

$$\boldsymbol{\iota} = \tfrac{1}{2}\partial_{\boldsymbol{\xi}}\Delta(\boldsymbol{\xi}).$$

Since nothing of this kind holds as a result of (7.44) when $\mathbf{T} \neq \mathbf{1}$, little is said about dissipation functions in more recent Onsagerist literature.

is the flux $\mathbf{h}$ or grad θ? Since neither is the time derivative of $\mathbf{F}$ or θ and since by $(7.45)_5$ the state is $(\mathbf{F}, \theta)$, the answer must be *neither*. This was the problem faced by ONSAGER and CASIMIR. They chose to divide the body up into little boxes, in each of which homogeneous conditions were claimed to prevail. The difference between the temperature of the box and the equilibrium temperature was regarded by them, apparently, as indistinguishable from a time derivative. DE GROOT & MAZUR†, on the contrary, by dropping surface integrals and neglecting terms they consider to be small, arrive at $\partial_t \varepsilon$ as the one and only flux, but they do not give any equation of state in which ε is an independent variable. Anyway, by *neither* process does the Fourier-Duhamel law (7.1) result: It has to be assumed *in addition* to the Onsagerist force-flux relations‡. Thus, even if the mathematics adduced were compelling, it could not rightly be interpreted as giving more than a necessary condition for the Onsagerist structure and the classical theory of heat conduction to be compatible with each other.

Attempts to treat diffusion in this way are of the same kind§.

The *practice* of Onsagerism does not face these difficulties¶. It starts

† Pp. 69–77 of their treatise, cited above. The treatment here derives from Section 5 of their paper, "On Onsager's relations in a magnetic field," *Physica*, **19**:961–970 (1953).

‡ Professor INGO MÜLLER has shown me a modified derivation in which an assumption of local influence is added to those of DE GROOT & MAZUR so as to infer the Fourier-Duhamel law. The basic objections remain, however.

§ S. R. DE GROOT & N. G. VAN KAMPEN, "On the derivation of reciprocal relations between irreversible processes," *Physica*, **21**:39–47 (1954); see Section 4. S. R. DE GROOT & P. MAZUR, "Extension of Onsager's theory of reciprocal relations, I," *Phys. Rev.*, (2) **94**:218–224 (1954), see Section 8. *Cf.* also the discussion of how "vector flows degenerate into scalars" on pp. 218 ff. of S. R. DE GROOT, *Thermodynamics of Irreversible Processes*, North Holland, Amsterdam, 1952.

¶ For evidence of the method, see Appendix 2 to this lecture. For a further sample of the results, consider Section 9 of the paper of 1954 by DE GROOT & MAZUR, cited in the previous footnote: "There are 15 reciprocal relations amongst the 36 viscosity coefficients. This leaves us with 21 independent coefficients. It can be noted that in isotropic media one has two independent viscosity coefficients, but no reciprocal relations." Curiously, the "reciprocal relations" these authors allege are just those which FALKENHAGEN, *op. cit. ante*, had asserted for a "triclinic crystal", being no more than necessary and sufficient conditions for existence of a potential for the stress.

On p. 710 of J. O. HIRSCHFELDER, C. F. CURTISS, & R. B. BIRD, *Molecular Theory of Gases and Liquids*, John Wiley & Sons, Inc., New York, 1954, revised 1964, irreversible thermodynamics is represented as yielding only still weaker results when applied to viscous fluids: reduction of the number of independent viscosities from 81 to 36. Thus, for these authors, the Onsager relations suffice merely to reduce (7.17) to the form $\mathbf{T} = \mathbf{\nabla D}$; hence, they say, the "simple physical interpretation" of the Onsager relations here is that the viscous stress vanishes in a rigid-body motion.

To obtain the classical reduction of the number of independent viscosities to two after their own fashion, the Onsagerists have had to invoke a further assumption, from an expression for the entropy production or caloric growth as a bilinear form:

$$\Delta = \sum_{i=1}^{m} J_i X_i, \tag{7.46}$$

obtained from some prior consideration of the problem. The J_i are named "fluxes", and the X_i are named "forces". That is,

$$\begin{aligned} \boldsymbol{\iota} &\equiv (J_1, J_2, \ldots, J_m), \\ \boldsymbol{\xi} &\equiv (X_1, X_2, \ldots, X_m). \end{aligned} \tag{7.47}$$

The phenomenological relations are then asserted in the form (7.27), and the Onsager relations in the form (7.44). While the Onsagerists usually deny that this is all they do, their own written record belies them. In their treatise De Groot & Mazur, despite having stated the condition (7.42) at the beginning and in their Chapter VI, use nothing but (7.46) and (7.47) in their Chapter XI, entitled "Heat Conduction, Diffusion, and Cross-effects", and in their Chapter XII, "Viscous Flow and Relaxation Phenomena". In these two chapters, which stretch over more than 100 pages, I can find no word of caution. The elaborate statistical treatment in their Chapter VII, apparently intended to persuade the reader that all is well both before and after, has just the opposite effect if really read through, for, even if the mathematics there be accepted without question, the analysis brings out the essential role of the conditions (7.40) and (7.42), *which do not hold* in the "applications" in the earlier chapters. The writers of a more "applied" slant do not even bother with the cautionary preface: For example, Donald Miller in his definitive paper in support of the Onsager relations writes, "The only significant addition of this century to the theory of irreversible thermodynamics . . . is the following: '*Provided the J_i and the X_i are chosen from the expression for Δ and are independent, the phenomenological coefficients Λ_{ij} of the linear laws satisfy the symmetry relation $\Lambda_{ij} = \Lambda_{ji}$.*' "

Significant as this "addition" may be, in this form it is nonsense.

"the Curie symmetry principle", which we shall discuss at the end of this lecture. In the treatise of De Groot & Mazur, the reader may have some difficulty in collecting the method from the remarks on viscosity on pp. 32–34 and 307–308 (referring back to an unspecified part of Chapter VI, presumably pp. 60–64), but he will see that in the end these authors *assume* the fluid to be isotropic. This assumption, as was shown by Cauchy and Stokes and may be seen in any good book on hydrodynamics, suffices by itself, with no Onsager relations and no Curie principle, to reduce the number of coefficients to two.

As we have remarked above, the principle of material frame-indifference forces all materials described by (7.17) to be isotropic fluids and hence to have at most two independent viscosities, in accord with hydrodynamics as hydrodynamicists have known it for over a century.

After all, J_i and X_i in (7.46) are merely letters. If we choose to set

$$\boldsymbol{\iota}' \equiv (X_1, J_2, \ldots, J_m), \qquad \boldsymbol{\xi}' \equiv (J_1, X_2, \ldots, X_m), \tag{7.48}$$

the linear relation that in terms of the choice (7.47) has the form $\boldsymbol{\iota} = \boldsymbol{\Lambda\xi}$, can be written equally well in the form

$$\boldsymbol{\iota}' = \boldsymbol{\Lambda}'\boldsymbol{\xi}', \tag{7.49}$$

and if $\mathbf{T}\boldsymbol{\Lambda}$ is symmetric, $\mathbf{T}\boldsymbol{\Lambda}'$ generally is not. That is, *symmetry or lack of it is meaningless unless we have some rule for selecting forces and fluxes*†. The Onsagerists give us such a rule, namely, (7.40) and (7.42), but, as it stands, *it does not apply* to heat conduction, viscosity, and diffusion, or, so far as I can see, to any problem governed by fields rather than by functions of time alone, since in these problems none of the entries in the bilinear form giving the caloric growth is the time derivative of an independent variable in the caloric equation of state.

As we have said, the Onsagerists try to escape from this difficulty by replacing the field equations by ordinary differential equations with constant coefficients and then effecting some kind of passage to the limit, as if all the field physics and mathematical theory of partial differential equations created in the past two centuries were a superfluous refinement! An alternative approach was explored by MEIXNER many years ago. MEIXNER was, in fact, the first to propose the Onsagerist theory in phenomenological form (1942). He asked, in effect, whether it really made any difference how the forces and fluxes are selected. He remarked‡ that if we define new fluxes and forces by non-singular linear transformation of the old ones,

$$\bar{\boldsymbol{\iota}} = \mathbf{A}\boldsymbol{\iota}, \qquad \bar{\boldsymbol{\xi}} = \mathbf{B}\boldsymbol{\xi}, \tag{7.50}$$

then in order that Δ be invariant,

$$\Delta = \boldsymbol{\iota} \cdot \boldsymbol{\xi} = \bar{\boldsymbol{\iota}} \cdot \bar{\boldsymbol{\xi}}, \tag{7.51}$$

† This fact has been remarked in various contexts and with various degrees of clarity, emphasis, and interpretation by:

J. E. VERSCHAFFELT, "Sur l'effet thermomécanique," *Bull. Classe Sci. Acad. Roy. Belg.*, **37**:853–872 (1951).

R. O. DAVIES, "Transformation properties of the Onsager relations," *Physica*, **18**:182 (1952).

B. D. COLEMAN & C. TRUESDELL, "On the reciprocal relations of Onsager," *J. Chem. Phys.*, **33**:28–31 (1960).

‡ J. MEIXNER, "Zur Thermodynamik der irreversiblen Prozessen in Gasen mit chemisch reagierenden, dissozierenden und anregbaren Komponenten," *Ann. Physik*, (5) **43**:244–270 (1943). The extension to take into account "even" and "odd" variables is given in a later paper by MEIXNER, "Consistency of the Onsager-Casimir reciprocal relations," in press.

it is necessary and sufficient that

$$\mathbf{B} = (\mathbf{A}^{-1})^{\mathrm{T}}. \tag{7.52}$$

When this condition is satisfied, the linear relation (7.27) is equivalent to

$$\bar{\boldsymbol{\iota}} = \bar{\boldsymbol{\Lambda}}\boldsymbol{\xi} \qquad \text{where } \bar{\boldsymbol{\Lambda}} = \mathbf{A}\boldsymbol{\Lambda}\mathbf{A}^{\mathrm{T}}. \tag{7.53}$$

In order that $\bar{\boldsymbol{\Lambda}}$ satisfy the same rule of symmetry as $\boldsymbol{\Lambda}$ does, namely, (7.44), it is necessary and sufficient that the particular linear transformations $\boldsymbol{\Lambda}$ and $\mathbf{A}$ satisfy the following time-reversal identity:

$$(\mathbf{TAT})\boldsymbol{\Lambda}(\mathbf{TAT})^{\mathrm{T}} = \mathbf{A}\boldsymbol{\Lambda}\mathbf{A}^{\mathrm{T}}. \tag{7.54}$$

A *sufficient* condition that (7.54) hold for all $\boldsymbol{\Lambda}$ is that $\mathbf{A}$ commute with time reversal:

$$\mathbf{TA} = \mathbf{AT}. \tag{7.55}$$

In this case, $\mathbf{T}\bar{\boldsymbol{\iota}} = \mathbf{A}(\mathbf{T}\boldsymbol{\iota})$ and $\mathbf{T}^{\mathrm{T}}\bar{\boldsymbol{\xi}} = \mathbf{B}(\mathbf{T}^{\mathrm{T}}\boldsymbol{\xi})$; that is, $\mathbf{T}\boldsymbol{\iota}$ and $\mathbf{T}^{\mathrm{T}}\boldsymbol{\xi}$ transform as do $\boldsymbol{\iota}$ and $\boldsymbol{\xi}$, respectively.† If $\mathbf{T} = \mathbf{1}$ (*i.e.*, all variables are "even", as was assumed in the older work), the condition (7.55) holds for all $\mathbf{A}$.

While some readers seem to interpret MEIXNER's theorem as a license to choose forces and fluxes at will, it is no such thing, nor did MEIXNER present it as such. Rather, it merely provides a class of equivalent statements of symmetry, once *one* choice of forces and fluxes has been made. It brings us no closer to that first choice of forces and fluxes, and hence no closer to a meaning for any particular statement of symmetry, than we were to start with. Indeed, it emphasizes the need for a definite choice, since it shows that without one, any claim of universal symmetry is simply false.

The Onsagerists have tried to escape from this dilemma. Some of them are content that their fluxes shall be time derivatives of *something*, but since

$$f(t) = \dot{g}(t) \qquad \text{where } g(t) = \int f(t)\, dt, \tag{7.56}$$

this requirement is empty‡. Others regard Onsager's theorem as a state-

† In his paper presently in press, MEIXNER generalizes these results so as to allow transformations of the type

$$\bar{\boldsymbol{\iota}} = \mathbf{A}\boldsymbol{\iota} + \mathbf{C}\boldsymbol{\xi}, \qquad \bar{\boldsymbol{\xi}} = \mathbf{B}\boldsymbol{\xi} + \mathbf{D}\boldsymbol{\iota}.$$

‡ *Cf.* the recent attempt at salvage by H. ZIEGLER, "A possible generalization of Onsager's theory," pp. 411–424 of *Proc. IUTAM Symposium Vienna, June 22–28, 1966*, Springer, Vienna, 1968. As a criterion ZIEGLER proposes, ". . . the time rates are the fluxes and their coefficients are the associated forces." Inspecting the product $-\mathbf{h} \cdot \boldsymbol{\gamma}$ in (7.24), he claims that "according to our criterion" the first factor is a flux

ment that by a "proper choice" of force and flux in (7.46), we can render $\mathbf{\Lambda}$ symmetric. This is true. Indeed, if with one choice we have $\Delta = \boldsymbol{\xi} \cdot \boldsymbol{\iota}$ and $\boldsymbol{\iota} = \mathbf{\Lambda}\boldsymbol{\xi}$, then, whatever be the symmetries of $\mathbf{\Lambda}$, if we set

$$\bar{\boldsymbol{\iota}} \equiv \boldsymbol{\iota} - \mathbf{\Lambda}_{-}\boldsymbol{\xi} = \mathbf{\Lambda}_{+}\boldsymbol{\xi}, \tag{7.57}$$

it follows that $\Delta = \boldsymbol{\xi} \cdot \bar{\boldsymbol{\iota}}$, so $\bar{\boldsymbol{\iota}}$ is a "proper choice" for getting a symmetric coefficient $\mathbf{\Lambda}_{+}$. Since this fact is no more than a trivial remark in algebra, it is scarcely a revolutionary discovery in physics. Still others ask what all the fuss is about. They say their physical intuition tells them which are the fluxes and which are the forces, so they need no mathematical rule. This statement amounts to an outright claim that by their physical intuition these blessed persons know which phenomenological coefficients satisfy symmetry relations. This may be true, but if it is, it knocks away all need for the Onsagerist *theory*, since it merely assures us that those endowed with sufficient intuition need no theory to help them in picking out variables such as to describe the response of every material by symmetric tensors alone.

There is a third approach, which circumvents the Onsagerist dilemma in regard to field phenomena. MAXWELL's kinetic theory of gases, based on integration of the Maxwell-Boltzmann differential equation, is fully specific. There is no need ever to mention "forces" and "fluxes" or even entropy; such symmetry relations as may be proved are definite and free from objection. Two heavy prices must be paid: First, the theory represents only an extremely special substance, and, second, it is of a high order of difficulty in mathematics. I will discuss these prices in turn.

The kinetic theory furnishes a special but illuminating model of irreversibility, motivated by considering the simplest kind of molecular structure. The main stochastic assumption embodies a type of fading memory not strictly compatible with the mechanics of systems of points subject to time-independent mutual forces: No matter what the past history of the molecular motions, at any given instant the pairs of molecules are statistically independent. The purely mechanical hypotheses are equally strong and special. First, the molecules are mathematical points; second, all collisions are binary even though the intermolecular forces may extend to ∞; third, stress and heating flux are taken to be manifestations of transfer of momentum and energy alone, leaving out of account such intermolecular forces as there may be. The *definitions* are so framed from

---

and the second is the associated force, but he does not reveal who or what directed him to apply (7.56) to $\mathbf{h}$ rather than to γ.

ZIEGLER does achieve a great advance relative to the other Onsagerist literature: He notices the papers in which the Onsagerist theory for fields is shown to be self-contradictory and concedes, "There is no doubt that these objections are serious."

the outset that in all conditions of the gas some very special local equations of state hold:

$$p = \frac{k}{m}\rho\theta, \qquad \varepsilon = \frac{3}{2}\frac{k}{m}\theta, \qquad \gamma = \tfrac{5}{3}; \tag{7.58}$$

the fluid is isotropic; the normal traction is never a tension:

$$\mathbf{n}\cdot\mathbf{Tn} \leqq 0 \qquad \text{for all } \mathbf{n}; \tag{7.59}$$

and if the Navier-Poisson-Stokes constitutive equation (2.31) is compatible with the theory in any sense, the Stokes relation must hold†:

$$3\lambda + 2\mu = 0. \tag{7.60}$$

These special features are not defects, for they are appropriate to the type of material the formal kinetic theory is designed to represent, namely, a moderately rarefied monatomic gas. However, when something is proved in the kinetic theory, we cannot be sure that it is a fact about fluids in general, let alone solids; on the contrary, it may just as well bring out some aspect of the peculiarly simple model laid down at the outset.

In an isotropic material, as we have seen, the Onsager relations asserted for heat conduction and viscosity are weaker than the requirements of material symmetry, and the kinetic gas is of course isotropic. Of the three classical phenomena we are discussing in this lecture, diffusion in a mixture of three or more constituents is the only one for which we can even address a question to the kinetic theory. In Onsagerist circles it seems to be widely believed that STEFAN's reciprocal relations,

$$f_{\mathfrak{ab}} = f_{\mathfrak{ba}}, \tag{7.22}$$

are verified by the kinetic theory. While this may be true, I cannot find a proof of it anywhere in the voluminous literature. The standard treatise, that by CHAPMAN & COWLING‡, does not mention it. HIRSCHFELDER, CURTISS, & BIRD in their book§, which upholds the cause of Onsagerism,

† That this result follows from the framing of the definitions rather than as a theorem proved from the Maxwell-Boltzmann equation was shown in my paper "On the viscosity of fluids according to the kinetic theory," *Z. Physik*, **131**:273–289 (1952), and is shown more clearly in the next lecture.

‡ S. CHAPMAN & T. G. COWLING, *The Mathematical Theory of Non-Uniform Gases*, Cambridge University Press, New York, 1939. Diffusion in mixture of more than two gases is not taken up specifically in the first edition of CHAPMAN & COWLING's book; in the second edition, 1957, it is mentioned in Note F in the Appendix. As we have seen, in a binary mixture (7.22) is a trivial consequence of the principle of linear momentum, so almost none of the older studies of the kinetic theory are relevant to the question.

§ See Section 11.2 of HIRSCHFELDER, CURTISS, & BIRD, *op. cit.*

go into the matter at length. They succeed in showing that for a ternary mixture, STEFAN's relations hold at the first stage of a certain formal expansion procedure, not the Chapman-Enskog procedure itself but a sub-procedure of it. Beyond the unproved validity of both these schemes of successive approximation, their work leaves open the question of whether reciprocity relations do in fact hold according to higher stages in their own procedures or merely express the nature of some first-order perturbation in some purely mathematical scheme. The same must be said of the arguments given by DE GROOT & MAZUR in the standard Onsagerist treatise†. As MÜLLER has remarked, from MAXWELL's calculations we may show that for Maxwellian molecules, STEFAN's relations hold exactly, but we cannot be sure that the result can be extended to other kinds of molecules‡. Today's chief expert on the kinetic theory, H. GRAD, has taken up the question. He has shown§ that a very broad class of symmetry relations holds according to the *linearized* Boltzmann equation, but his analysis leaves open the question for the precise kinetic theory. He has informed me, moreover, that unpublished calculations by one of his students fail to confirm Stefan's relations for a mixture of gases whose molecules are rough spheres instead of points.

Now the Onsager relations, according to their supporters, are statements about *all* irreversible processes in *all* materials, provided the constitutive equations be linear. Thus a proof of such relations in the kinetic theory, in order to be relevant, must hold for *all* molecular models. If there is a *single* model for which the asserted rule of symmetry fails, "Onsager's theorem" is proved false, and we are thrown back to a statement like this: A simpler molecular structure yields simpler phenomenological coefficients. This is little more than a platitude.

† Chapter IX, Section 7, of their treatise, cited above. There the assumption is concealed by the phrase, "For simplicity's sake we shall now consider the case that the temperature T and the average velocity $\mathbf{v}$ are uniform." Since according to ENSKOG's first approximation the distribution function depends upon grad$T$ and grad$\mathbf{v}$ as parameters, so do the diffusion coefficients. Thus DE GROOT & MAZUR's treatment also rests on an assumption, not explained as such, of quasi-equilibrium, and their argument, to the extent it can be accepted at all, establishes STEFAN's relations only for the equilibrium values of the $f_{\mathfrak{ab}}$.

‡ We recall that some entire effects possible in the general kinetic gas disappear for the inverse fifth-power molecule. For example, the coefficient of the contribution to the stress by the products of pressure gradients is 0 for Maxwellian molecules although for other models it is not even small compared to the coefficient of the similar term for the temperature gradients. Specifically, on pp. 268–269 CHAPMAN & COWLING give for Maxwellian molecules $\varpi_4 = 0$ and $\varpi_5 = 3$, while for rigid spheres $\varpi_4 = 0.68$ and $\varpi_5 = 0.22$, showing how sensitive the phenomenological coefficients are to the choice of intermolecular force.

§ H. GRAD, "Theory of rarefied gases," *Rarefied Gas Dynamics*, Pergamon, London, 1960. See pp. 111 and 125.

Because of their concreteness, the results from the kinetic theory suggest a positive rather than merely critical approach to the Onsager relations. First, we come out of the fog of "forces" and "fluxes" and ask specific questions about specific quantities having clear physical meanings. In the present instance, the question is, do Stefan's reciprocal relations (7.22) hold in general, and if not, when do they hold? The results in the kinetic theory suggest to me that Stefan's relations for diffusion, like the Stokes relation for the viscosities, may well be not general but rather appropriate to conditions such as hold only approximately and in some cases. The kinetic theory itself, by its nature, is restricted to dilute gases and to binary encounters of molecules. Even so, it does not generally yield merely binary interaction between the *constituent gases* of a multiple mixture; however, most of the mathematical schemes of approximation are so arranged as to start by considering only binary interaction at the first stage. This idea suggested to me, some years ago†, a phenomenological definition of a special kind of mixture such as to obey Stefan's relations. We may say that a mixture is *dilute* when its diffusive drag coefficients for the pairs depend on the densities of the pairs only:

$$f_{\mathfrak{ab}} = f_{\mathfrak{ab}}(\rho_{\mathfrak{a}}, \rho_{\mathfrak{b}}), \qquad f_{\mathfrak{ab}}(0, \rho_{\mathfrak{b}}) = 0, \tag{7.61}$$

rather than $f_{\mathfrak{ab}} = f_{\mathfrak{ab}}(\rho_1, \rho_2, \ldots, \rho_{\mathfrak{f}})$. That is, the diffusive forces are binary forces, and at a place where one constituent is absent, it experiences no drag in virtue of diffusion. It is easy to show then that the principle of conservation of total linear momentum, which we have already expressed in the form

$$\sum_{\mathfrak{a}=1}^{\mathfrak{f}} (f_{\mathfrak{ab}} - f_{\mathfrak{ab}}) = 0, \qquad \mathfrak{b} = 1, 2, \ldots, \mathfrak{f}, \tag{7.23}$$

requires that

$$f_{\mathfrak{ba}} = f_{\mathfrak{ab}}. \tag{7.22}$$

That is, *in a dilute mixture Stefan's reciprocal relations are equivalent to the conservation of linear momentum.* Thus these particular "Onsager relations" join the symmetry of the heat-conductivity tensor as *necessary consequences of certain material symmetries.* In particular, according to the kinetic theory the diffusive drags do satisfy (7.61) if the molecules are Maxwellian; whether or not they do for other molecules, is not presently known. To find a mixture in which the Onsager relations do not hold in virtue of general physical principles alone, we have to investigate those in which the diffusive drags are not binary. A non-Maxwellian mixture

† *Cf.* the two papers of mine cited on p. 117; see p. 167 of the former, Section 6 of the latter.

according to the kinetic theory may furnish an example if treated rigorously, but here the problem remains open.

While at first the Onsager relations were regarded as the A and Ω of "irreversible thermodynamics", in time need began to be felt for some further idea. Consider, for example, a fluid endowed with both heating flux **h** and viscous stress **V**. Then

$$\Delta = \operatorname{tr}(\mathbf{VD}) - \boldsymbol{\gamma}\cdot\mathbf{h}. \tag{7.62}$$

A good Onsagerist knows what the fluxes are; he has already picked them for viscosity and heat conduction alone, namely, the components of **D** and **h**, respectively. Thus for both phenomena together he takes the nine-dimensional vectors

$$\begin{aligned} \boldsymbol{\iota} &= (D_{11}, \sqrt{2}\,D_{12}, \ldots, D_{33}, h_1, h_2, h_3), \\ \boldsymbol{\xi} &= (V_{11}, \sqrt{2}\,V_{12}, \ldots, V_{33}, -\gamma_1, -\gamma_2, -\gamma_3), \end{aligned} \tag{7.63}$$

so (7.26) is satisfied. According to Onsagerist principles, the phenomenological relations are (7.27), namely, in the present case

$$\begin{aligned} D_{11} &= \sum_{i=1}^{3} \Lambda_{1j}V_{jj} + \sqrt{2}\,(\Lambda_{14}V_{12} + \Lambda_{15}V_{23} + \Lambda_{16}V_{31}) - \sum_{j=7}^{9} \Lambda_{1j}\gamma_j, \\ &\cdot \\ &\cdot \\ &\cdot \\ h_3 &= \sum_{i=1}^{3} \Lambda_{9j}V_{jj} + \sqrt{2}\,(\Lambda_{94}V_{12} + \Lambda_{95}V_{23} + \Lambda_{96}V_{31}) - \sum_{j=7}^{9} \Lambda_{9j}\gamma_j, \end{aligned} \tag{7.64}$$

where, by ONSAGER, the nine-dimensional tensor $\boldsymbol{\Lambda}$ is symmetric. Now our Onsagerist knows that symmetry of $\boldsymbol{\Lambda}$ is not enough: For agreement with the classical theories of FOURIER, NAVIER, POISSON, and STOKES, which he does not question, he wishes (7.64) for a fluid to reduce to *two separate constitutive equations*, viz

$$\mathbf{V} = \lambda(\operatorname{tr}\mathbf{D})\mathbf{1} + 2\mu\mathbf{D}, \tag{2.33}$$
$$\mathbf{h} = K\boldsymbol{\gamma}. \tag{7.65}$$

To this end, he has recourse to what is sometimes regarded as the second fundamental principle of irreversible thermodynamics, which he calls "CURIE's principle", "CURIE's theorem", or "CURIE's axiom", according to taste.

So far as I can learn, the first occurrence of this nostrum is to be found in a paper published in 1951 by PRIGOGINE & MAZUR†: " . . . all

† I. PRIGOGINE & P. MAZUR, "Sur deux formulations de l'hydrodynamique et le problème de l'hélium liquide, II," *Physica*, **17**:661–679 (1951). See Section 3d.

coupling between quantities of different tensorial character being forbidden . . . ," but CURIE is not cited as the author of this Olympian interdiction. KIRKWOOD & CRAWFORD† were quick to follow suit: "We must treat scalars, vectors and tensors separately, for entities of different tensorial character cannot interact (CURIE's theorem)." In the first general treatise on Onsagerism, DE GROOT‡ gave at least four different statements, of which I quote here two: "It is impossible for a force of a certain tensorial character to give rise to a flow of a different tensorial character (CURIE's theorem)," and CURIE's theorem "forbids combinations of vectors and scalars." The Onsagerist writers seem with time to have smelled something rotten in the (thermodynamic) state of the Low Countries and have become more cautious. Perhaps they noticed that the scalar tr $\mathbf{A}$ is a linear isotropic function of the tensor $\mathbf{A}$, which seems to allow a pretty active interaction if not a licentious combination; thus HIRSCHFELDER, CURTISS, & BIRD§ write, "However, if the system is isotropic, it may be shown that those terms which correspond to a coupling of tensors whose orders differ by an odd number do not occur." In their recent treatise DE GROOT & MAZUR¶ state the "Curie symmetry principle" still more guardedly: "As a consequence symmetry properties of the material system considered may have the effect that the components of the fluxes do not depend on all components of the thermodynamic forces . . . . Thus, in particular for an isotropic system . . . it can be shown that fluxes and thermodynamic forces of different tensorial character do not couple," and they devote some pages to arguments adduced in favor of their proclamation of prohibited degrees of consanguinity.

While some of these statements seem to throw us back to that desperation point of deciding what quantities are fluxes, a little experience in the Onsagerist literature shows that the authors are groping toward a purely algebraic theorem. What this theorem is, we may have some difficulty in divining, since the terms "interact" and "couple" do not occur in algebra and are never defined by Onsagerists, although they do occur frequently in *The Arabian Nights.* To every statement I have seen‖, Fourier's law for isotropic materials with a linear conductivity $K(\theta)$ seems to be a counterexample:

$$\mathbf{h} = (K_0 + K_1\theta) \text{ grad } \theta, \tag{7.66}$$

† J. G. KIRKWOOD & B. CRAWFORD, JR., "The macroscopic equations of transport," *J. Phys. Chem.*, **56**:1048–1051 (1952). See p. 1050.

‡ S. R. DE GROOT, *Thermodynamics of Irreversible Processes*, North Holland, Amsterdam, 1952. See pp. 77, 100, 121, 192.

§ *Op. cit.*, p. 709.

¶ *Op. cit.*, pp. 5, 31–34, 57–64.

‖ Except the first of the two just quoted from the treatise of DE GROOT & MAZUR, which the conditional "may" renders empty and hence not disprovable.

since the scalar θ seems to be "interacting" or "coupling" with the vectors **h** and grad θ. Moreover, in a typical application, CURIE's theorem is invoked so as to infer that if

$$\alpha + \mathbf{a} \cdot \mathbf{b} + \operatorname{tr} (\mathbf{A}\mathbf{B}^{\mathrm{T}}) \geqq 0, \tag{7.67}$$

where α, **a**, and **A** are functions of **b** and **B**, then

$$\alpha \geqq 0, \qquad \mathbf{a} \cdot \mathbf{b} \geqq 0, \qquad \operatorname{tr} (\mathbf{A}\mathbf{B}^{\mathrm{T}}) \geqq 0. \tag{7.68}$$

Of course, this conclusion is wrong unless restrictions be imposed—typically, that the functions α, **a**, and **A** be linear. Otherwise, there are trivial counterexamples, *e.g.*,

$$\alpha = - \operatorname{tr} (\mathbf{B}\mathbf{B}^{\mathrm{T}}), \qquad \mathbf{a} = \mathbf{0}, \qquad \mathbf{A} = 2\mathbf{B}. \tag{7.69}$$

I cannot resist the temptation to flash on the screen† the passage in CURIE's works the Onsagerists cite for their theorem and offer a free copy of *The Non-Linear Field Theories of Mechanics* as a penalty to anyone who can find there any of the statements used by Onsagerists. At least, we can recognize the word "liaison". The burden of the paper by CURIE is to express belief that the symmetries of different constitutive functions for the same material are related, *e.g.*, we expect an optically isotropic material to be thermally isotropic as well, but neither CURIE nor anyone else can prove this is always so. In his conclusion, CURIE tries to draw profit from the ideas that "There is no effect without a cause" and that "There is no cause without an effect," but this type of physics is not in vogue today and cannot be blamed on the Onsagerists.

However, I will not make much of these obscurities, for they, unlike those of ONSAGER's reciprocity relations, are easy to clear away. The reader of any good book on crystal physics knows that every constitutive equation can be given a form such as to render assigned symmetries automatic. This idea goes back to CAUCHY and was illustrated in brilliant examples before CURIE was born. In modern continuum mechanics, which began to develop at the same time as Onsagerism and the existence of which the Onsagerists take care not to notice, many such reductions have been made explicit and precise. For example, the three functional relations

$$\begin{aligned} B &= F(\mathbf{A}, \mathrm{a}), \\ \mathbf{b} &= \mathbf{f}(\mathbf{A}, \mathrm{a}), \\ \mathbf{B} &= \mathbf{F}(\mathbf{A}), \end{aligned} \tag{7.70}$$

† These pages are reproduced here as Figure 8. They are from the paper of P. CURIE, "Sur la symétrie dans les phénomènes physiques, symétrie d'un champ électrique et d'un champ magnétique," *J. Physique Radium*, (3)**3**, 393–415(1894); *Oeuvres*, Gauthier-Villars, Paris, 1908, pp. 118–141. While sometimes the Onsagerists cite p. 129 of the latter printing, I can find there nothing even vaguely related to interactions of any kind.

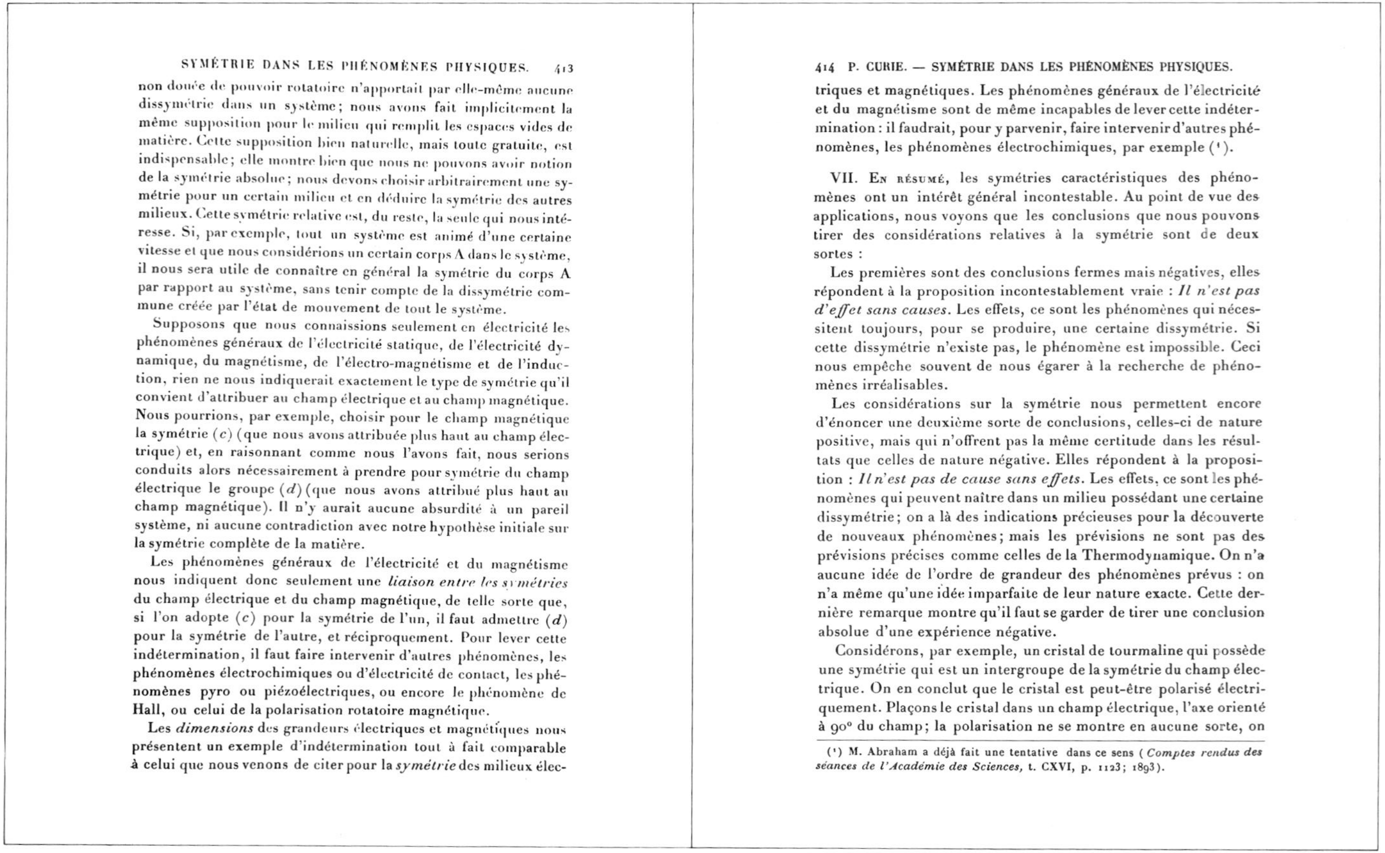
SYMÉTRIE DANS LES PHÉNOMÈNES PHYSIQUES. 413

non douée de pouvoir rotatoire n'apportait par elle-même aucune dissymétrie dans un système; nous avons fait implicitement la même supposition pour le milieu qui remplit les espaces vides de matière. Cette supposition bien naturelle, mais toute gratuite, est indispensable; elle montre bien que nous ne pouvons avoir notion de la symétrie absolue; nous devons choisir arbitrairement une symétrie pour un certain milieu et en déduire la symétrie des autres milieux. Cette symétrie relative est, du reste, la seule qui nous intéresse. Si, par exemple, tout un système est animé d'une certaine vitesse et que nous considérions un certain corps A dans le système, il nous sera utile de connaître en général la symétrie du corps A par rapport au système, sans tenir compte de la dissymétrie commune créée par l'état de mouvement de tout le système.

Supposons que nous connaissions seulement en électricité les phénomènes généraux de l'électricité statique, de l'électricité dynamique, du magnétisme, de l'électro-magnétisme et de l'induction, rien ne nous indiquerait exactement le type de symétrie qu'il convient d'attribuer au champ électrique et au champ magnétique. Nous pourrions, par exemple, choisir pour le champ magnétique la symétrie (c) (que nous avons attribuée plus haut au champ électrique) et, en raisonnant comme nous l'avons fait, nous serions conduits alors nécessairement à prendre pour symétrie du champ électrique le groupe (d) (que nous avons attribué plus haut au champ magnétique). Il n'y aurait aucune absurdité à un pareil système, ni aucune contradiction avec notre hypothèse initiale sur la symétrie complète de la matière.

Les phénomènes généraux de l'électricité et du magnétisme nous indiquent donc seulement une *liaison entre les symétries* du champ électrique et du champ magnétique, de telle sorte que, si l'on adopte (c) pour la symétrie de l'un, il faut admettre (d) pour la symétrie de l'autre, et réciproquement. Pour lever cette indétermination, il faut faire intervenir d'autres phénomènes, les phénomènes électrochimiques ou d'électricité de contact, les phénomènes pyro ou piézoélectriques, ou encore le phénomène de Hall, ou celui de la polarisation rotatoire magnétique.

Les *dimensions* des grandeurs électriques et magnétiques nous présentent un exemple d'indétermination tout à fait comparable à celui que nous venons de citer pour la *symétrie* des milieux élec-

414 P. CURIE. — SYMÉTRIE DANS LES PHÉNOMÈNES PHYSIQUES.

triques et magnétiques. Les phénomènes généraux de l'électricité et du magnétisme sont de même incapables de lever cette indétermination : il faudrait, pour y parvenir, faire intervenir d'autres phénomènes, les phénomènes électrochimiques, par exemple (¹).

VII. En résumé, les symétries caractéristiques des phénomènes ont un intérêt général incontestable. Au point de vue des applications, nous voyons que les conclusions que nous pouvons tirer des considérations relatives à la symétrie sont de deux sortes :

Les premières sont des conclusions fermes mais négatives, elles répondent à la proposition incontestablement vraie : *Il n'est pas d'effet sans causes*. Les effets, ce sont les phénomènes qui nécessitent toujours, pour se produire, une certaine dissymétrie. Si cette dissymétrie n'existe pas, le phénomène est impossible. Ceci nous empêche souvent de nous égarer à la recherche de phénomènes irréalisables.

Les considérations sur la symétrie nous permettent encore d'énoncer une deuxième sorte de conclusions, celles-ci de nature positive, mais qui n'offrent pas la même certitude dans les résultats que celles de nature négative. Elles répondent à la proposition : *Il n'est pas de cause sans effets*. Les effets, ce sont les phénomènes qui peuvent naître dans un milieu possédant une certaine dissymétrie; on a là des indications précieuses pour la découverte de nouveaux phénomènes; mais les prévisions ne sont pas des prévisions précises comme celles de la Thermodynamique. On n'a aucune idée de l'ordre de grandeur des phénomènes prévus : on n'a même qu'une idée imparfaite de leur nature exacte. Cette dernière remarque montre qu'il faut se garder de tirer une conclusion absolue d'une expérience négative.

Considérons, par exemple, un cristal de tourmaline qui possède une symétrie qui est un intergroupe de la symétrie du champ électrique. On en conclut que le cristal est peut-être polarisé électriquement. Plaçons le cristal dans un champ électrique, l'axe orienté à 90° du champ; la polarisation ne se montre en aucune sorte, on

(¹) M. Abraham a déjà fait une tentative dans ce sens (*Comptes rendus des séances de l'Académie des Sciences*, t. CXVI, p. 1123; 1893).

FIGURE 8. CURIE's remarks on the connections among symmetries.

where B is a scalar, $\mathbf{a}$ and $\mathbf{b}$ are vectors, and $\mathbf{A}$ and $\mathbf{B}$ are symmetric tensors, are called *isotropic* if these functional equations hold for all orthogonal tensors $\mathbf{Q}$:

$$\begin{aligned} F(\mathbf{QAQ}^{\mathrm{T}}, \mathbf{Qa}) &= F(\mathbf{A}, \mathbf{a}), \\ \mathbf{f}(\mathbf{QAQ}^{\mathrm{T}}, \mathbf{Qa}) &= \mathbf{Qf}(\mathbf{A}, \mathbf{a}), \\ \mathbf{F}(\mathbf{QAQ}^{\mathrm{T}}) &= \mathbf{QF}(\mathbf{A})\mathbf{Q}^{\mathrm{T}}, \end{aligned} \tag{7.71}$$

respectively. According to theorems of algebra, the functions F, $\mathbf{f}$, and $\mathbf{F}$ are isotropic if and only if

$$\begin{aligned} F(\mathbf{A}, \mathbf{a}) &= g(\mathrm{tr}\,\mathbf{A}, \mathrm{tr}\,\mathbf{A}^2, \mathrm{tr}\,\mathbf{A}^3, a^2, \mathbf{a}\cdot\mathbf{Aa}, \mathbf{a}\cdot\mathbf{A}^2\mathbf{a}), \\ \mathbf{f}(\mathbf{A}, \mathbf{a}) &= (F_1\mathbf{1} + F_2\mathbf{A} + F_3\mathbf{A}^2)\mathbf{a}, \\ \mathbf{F}(\mathbf{A}) &= H_1\mathbf{1} + H_2\mathbf{A} + H_3\mathbf{A}^2, \end{aligned} \tag{7.72}$$

where g is a function, where F_1, F_2, F_3 are isotropic scalar functions of $\mathbf{A}$ and $\mathbf{a}$, and where H_1, H_2, H_3 are isotropic scalar functions of $\mathbf{A}$. These are the general solutions of the functional equations (7.71). The special cases that follow when the functions are linear may be read off by inspection. All this and much more we know from algebra.

As for concluding that the inequality $\Delta \geqq 0$ requires certain summands in Δ to be individually non-negative, this is harder because it is usually untrue. Modern continuum mechanics contains arguments of this kind, applied, with care, to special constitutive equations, as I have illustrated in Lecture 3. Explicit cases are known, *e.g.*, that of a visco-elastic solid that conducts heat, in which such a separation is *not* justified.

We may now summarize what we have learned about Onsagerism.

1. *Alleged basis in statistical mechanics.* Here the "fluxes" are clearly defined as expectations, but the phenomenological relations are imposed on the theory rather than derived from it, and in particular, extramechanical assumptions about stochastic processes are essential, even if the wobbly mathematics be let pass unchallenged.
2. *Basis in the kinetic theory of gases.* For the very special model of irreversibility afforded by the Maxwellian kinetic theory, again the "fluxes" are clearly defined, but those of the Onsager relations that are not trivial have never been proved to hold except in uniform or dilute conditions, or for Maxwellian molecules, and there are indications that they may not hold generally.
3. *Phenomenological theory based on taking the flux as the time-rate of change of the state.* Such a flux is well defined, and the Onsager relations (7.44) may be imposed as a self-consistent postulate. All attempts so far to apply this theory to heat con-

duction, viscosity, and diffusion rest on strong further assumptions *ad hoc* and on questionable mathematics.

4. *Phenomenological Onsagerism in practice.* Here fluxes are chosen by inspection of the entropy production. No self-consistent rule for doing so has ever been stated, and the "applications" are divined, not proved.
5. *Curie's theorem.* While no such thing as "CURIE's theorem" exists†, its apparent purpose is served by various known theorems of algebra which express material symmetries in reduced forms.

Here we find the final estimate. Appeal to an all-inclusive, vaguely surmised algebraic principle that has had to be patched up and tied back together from paper to paper is a symptom of the Onsagerists' steadfast refusal to commit themselves to anything specific in physics. The same thing has happened with the claimed reciprocity relations: First, symmetry was universal, then adjustment had to be made for "even" and "odd" variables‡, and finally the magnetic field had to be given a special role§, which I have not discussed in this lecture. While there is nothing in their formalism and expressed prescriptions to stop the Onsagerists from taking the magnetic field as being a state variable (an "even" one, of course), they do not do so, because they know that if they did, their rules would lead them into error. The magnetic field penetrated their citadel of thermodynamic universality and forced them to face it, specifically—*the magnetic field*, not just some α or X or J which contributes entropy. I suggest that the floating malaise outsiders sense in reading the Onsagerist literature can be cured by facing the physics frankly. It is gross physics the Onsagerists claim to study. Gross physics concerns, not X's and J's identified by no more than their roles in production of entropy, but a short list of specific set-functions and fields:

mass, motions of constituent bodies,
linear momentum, rotational momentum,
force, energy,

† Of course, this is the "non-existent theorem of algebra" mentioned on p. 50 of my *Six Lectures on Modern Natural Philosophy*, Springer-Verlag, New York, 1966.

‡ While this adjustment was first made in the paper by CASIMIR (1945), *op. cit.*, the Onsagerist literature ignored it long thereafter; it is not mentioned in the treatises by DE GROOT (1952) and by HIRSCHFELDER, CURTISS, & BIRD (1954, revised 1964).

§ P. MAZUR & S. R. DE GROOT, "On Onsager's relations in a magnetic field," *Physica*, **19**:961–970 (1953).

R. FIESCHI, S. R. DE GROOT, & P. MAZUR, "Thermodynamical theory of galvanometric and thermomagnetic phenomena. I, II, III," *Physica*, **20**:67–76, 245–248, 259–273 (1954).

temperature, calory (entropy),
charge-current, magnetic flux.

For these fields certain specific properties, representing physical laws, are laid down as axioms. It is *just these fields and functions*, not vague universals like X's and J's, that gross physics concerns. The Onsagerists' claimed generality is superfluous either for concrete application to cases or for mastery of basic principle. If the Onsager relations are to take on meaning, they must be referred to *specific* physical quantities. In this spirit, I have tried today to pull the discourse down from the smoggy sky of forces and fluxes by discussing the specific phenomena of heat conduction, viscosity, and diffusion.

In this same spirit, the Onsager relations may be saved from the growing belief, often voiced but rarely put in print, that they are nonsense. To me, the symmetry relations now known to exist in the continuum theories of certain particular materials, viewed against the nature of the assumptions made in the alleged proofs from statistical mechanics and kinetic theory, suggest that counterparts of what are claimed to be Onsager relations can be found and given a specific meaning in every *concrete* case, and that meaning will express on the phenomenological scale a *special symmetry*† or *stability* like the Stokes relation or isotropy or the Gibbs inequalities, and on the molecular scale a particularly simple, *special molecular structure* or a condition of *diluteness or uniformity*, or both.

I am sorry this has been such a long lecture: It takes more time to scotch falsehood and expose fables than it does to set forth something solid and new. I must confess, however, to having given the subject more time than it is worth.

Even if the Onsagerist thermodynamics were above criticism in its principles and details, it would still have to face an appraisal of its results.

First, it begins and ends with statements that certain matrices are or are not symmetric. What help is this? The number of experiments needed to verify some particular linear (and hence limited) theory of materials and to determine the coefficients in some constitutive equation is reduced a little, but what then comes out? According to the treatise of De Groot & Mazur‡, we are to obtain "phenomenological equations . . . complete in the sense that one now has at one's disposal a consistent set of partial differential equations for the state parameters of a material system, which

† *Cf.* the proof by Coleman that in a material with fading memory, the difference between the *initial* value and the equilibrium value of the stress-relaxation modulus, in the sense of linear visco-elasticity, is positive-semidefinite, and the *initial* value is symmetric, though later values need not be. See Section 8 of B. D. Coleman, "On thermodynamics, strain, impulses, and viscoelasticity," *Arch. Rational Mech. Anal.*, **17**:230–254 (1964).

‡ P. 4 of the treatise cited above.

may be solved with the proper initial and boundary conditions." Search of the book, however, does not provide support for this claim. While many partial differential equations are written down, most of these come from classical researches on continuum mechanics done in the last century, before the birth of any Onsagerist. With all this material taken from older domains, the further knowledge, if knowledge it be, that some matrix is symmetric, is not an overwhelming addition†. At that, I have not been able to find in the whole book a single worked example involving more than one independent variable except in acoustics, where the equations are linear with constant coefficients. In a recent survey‡ where he attempts to decide whether Onsagerism has produced "anything other than journal papers," JAMES WEI writes, "Engineers are, of course, quite familiar with linear models and used them for years before the emergence of Irreversible Thermodynamics. They should use the linear model provided by Irreversible Thermodynamics only if there is something to be gained," but he is unable to suggest a specific case of usefulness.

Second, Onsagerism claims to unify and correlate much existing knowledge of irreversible processes. As far as concerns heat conduction, viscosity, and diffusion, we have seen that this is not so. Not only does Onsagerism not apply to any of these phenomena without a Procrustean force-fit, but even in the generous interpretation of its sectaries it does not yield as much reduction for the theory of viscosity as was known a century earlier and follows from fundamental principles; such of its results for heat conduction as were not long previously known are suspect; and its results for diffusion are not even certainly supported in the kinetic theory of dilute monatomic gases. True, the reciprocity relations are not *forbidden* by any general principle, but this is faint praise indeed. In speaking of applications, WEI wrote, " . . . we do not count it as an application when someone shows that Irreversible Thermodynamics is consistent with what we already know. Ask not what we can do for Irreversible Thermodynamics; ask what Irreversible Thermodynamics can do for us."

Today, against Onsagerism there are heavy attacks on two fronts: from those who wish an applied or at least applicable science and are not satisfied by a bare list of cases in which matrices are symmetric, swelled by

† The whole Onsagerist controversy may well go the way of similar earlier ones: multiconstant or rariconstant elasticity, the Stokes relation in viscosity, the "kinetic" theory (*i.e.*, the RIVLIN-KUBO "neo-Hookean" stored-energy function) of rubber, WEISSENBERG's assertion in non-linear viscosity. In all these cases, the gain made if the coefficients do reduce as claimed is little or nothing in the solution of particular examples or in proof of general theorems, and the theorists may safely forget the matter and leave to experimenters particular values, reduced or not, for particular materials.

‡ J. WEI, "Irreversible thermodynamics in engineering," *Ind. Eng. Chem.*, **58**:55–60 (1966).

pages of transformations of the result into different notations with alternative variables, and from those who wish to understand the thermodynamics of deformation as a basic aspect of gross physics. The former, as WEI has said with much force, have to add so much to Onsagerism that they can pretty well get on without it. The latter find it a complicated, verbose formalism with uncertain meaning, no meat, no fruit. After the concrete thermodynamics of Lectures 1 to 6, where non-trivial problems involving non-linear field relations are set and solved mathematically so as to yield specific, detailed results fit for comparison with experiments on real bodies occupying limited portions of space, it is difficult to present Onsagerism except as comic relief. It is not that non-linearity is an end in itself or even that it is always necessary. Linear theories have wide application. The trouble here is that the Onsagerists use linearity, not as a specially simple and useful example of general principles, but as a shield to keep from having to face the basic issues of gross physics.

However, I do not wish to stop after no more than condemnation. The Onsagerist school, and particularly J. MEIXNER, undertook and presented an earnest, sincere study of the phenomena giving rise to energy losses in many parts of physics and its theory. While they did not find physical principles sufficient to describe and direct them, or a mathematical framework adequate to get on with the work, they deserve respect and credit for having drawn attention to this major field of science in the 1940s and early 1950s, when it was neglected by everyone else. In particular, their literature served as an impetus (if, indeed, mainly by reaction) toward the great new thermodynamics developed by COLEMAN and his associates.

No such estimate, however, can be just if it fails to cast in relief the still greater deserts of another early author, C. ECKART (1940), who obtained and used a major case of the Clausius-Duhem inequality, who treated heat conduction and viscosity explicitly on the basis of it, who attempted to split inequalities into parts without appeal to any non-existent theorem, who obtained some of the correct consequences of the irreversibility principle at an equilibrium state—and who did not obfuscate the scene by any circular or inapplicable rule of symmetry.

Note

This lecture, the longest in the series, is intended to replace and document pages 49–59 of *Six Lectures on Modern Natural Philosophy* (Springer-Verlag, New York, 1966), which the reactions of some readers show to have been too concise.

APPENDIX 1

ON THE SYMMETRY OF THE HEAT-CONDUCTION TENSOR

BY C.-C. WANG

For a material satisfying Fourier's law of heat conduction, the heating flux $\mathbf{h}$ depends linearly on the temperature gradient:

$$\mathbf{h} = \mathbf{K} \operatorname{grad} \theta, \tag{7A.1}$$

where the tensor $\mathbf{K}$ is the *thermal conductivity*. In general, $\mathbf{K}$ is a function of the deformation gradient $\mathbf{F}$ and the temperature θ:

$$\mathbf{K} = \mathbf{K}(\mathbf{F}, \theta) \tag{7A.2}$$

for each fixed material element.

In the linear "thermodynamics of irreversible processes", it is often claimed that $\mathbf{K}$ must be a symmetric tensor, this condition being one of the so-called *Onsager relations*. As explained in the lecture preceding, no firm theoretical basis has been found for this claim.

Earlier, STOKES (1851) remarked that for a number of crystal classes, the *principle of material symmetry* sufficed to prove that $\mathbf{K}$ is symmetric. The purpose of this appendix is to amplify his remark by an exhaustive analysis, resting upon the basic assumption of the *principle of material frame-indifference*, which STOKES used only implicitly. To simplify the argument, we assume that $\mathbf{K}$ is non-singular.

According to the principle of frame-indifference, for every non-singular $\mathbf{F}$ the tensor function $\mathbf{K}$ satisfies the identity

$$\mathbf{K}(\mathbf{QF}, \theta) = \mathbf{QK}(\mathbf{F}, \theta)\mathbf{Q}^T \tag{7A.3}$$

if $\mathbf{Q} \in \mathscr{o}$, where $\mathscr{o}$ is the full orthogonal group in three dimensions. The *symmetry group* $\mathscr{g}_\mathbf{N}$ of the material relative to the fixed reference configuration $\mathbf{N}$ is the set of unimodular tensors $\mathbf{H}$ such that

$$\mathbf{K}(\mathbf{FH}, \theta) = \mathbf{K}(\mathbf{F}, \theta) \tag{7A.4}$$

for all non-singular tensors $\mathbf{F}$. The reference configuration $\mathbf{N}$ is that with respect to which $\mathbf{F}$ is calculated. Thus $\mathbf{F} = \mathbf{1}$ in the reference configuration. We shall denote the value of $\mathbf{K}$ there by $\mathbf{K}_0$:

$$\mathbf{K}_0 = \mathbf{K}(\mathbf{1}, \theta). \tag{7A.5}$$

From (7A.4), also,

$$\mathbf{K}_0 = \mathbf{K}(\mathbf{H}, \theta) \qquad \text{if } \mathbf{H} \in \mathcal{g}_{\mathbf{N}}. \tag{7A.6}$$

According to NOLL (1958), a material is a *solid* if there exists some reference configuration $\mathbf{N}$ such that

$$\mathcal{g}_{\mathbf{N}} \subset \mathcal{o}. \tag{7A.7}$$

Such a reference configuration $\mathbf{N}$ is called *undistorted*. We are interested in finding the restrictions on the conductivity $\mathbf{K}_0$ at an undistorted state of a solid.

Since $\mathcal{g}_{\mathbf{N}}$ satisfies (7A.7), every member $\mathbf{H}$ of $\mathcal{g}_{\mathbf{N}}$ may be substituted for $\mathbf{Q}$ in (7A.3). Thus, in particular,

$$\mathbf{H}\mathbf{K}_0\mathbf{H}^T = \mathbf{K}(\mathbf{H}, \theta). \tag{7A.8}$$

Consequently, by (7A.6),

$$\mathbf{H}\mathbf{K}_0 = \mathbf{K}_0\mathbf{H}. \tag{7A.9}$$

That is, *every member of the symmetry group commutes with* $\mathbf{K}_0$. This condition allows us to determine the restrictions imposed on $\mathbf{K}_0$ for various symmetry types appropriate to solids.

The tensor $\mathbf{K}_0$ has a unique polar decomposition of the form

$$\mathbf{K}_0 = \mathbf{R}\mathbf{U}, \tag{7A.10}$$

where $\mathbf{R}$ is orthogonal and $\mathbf{U}$ is positive-definite and symmetric. By (7A.9),

$$\mathbf{H}\mathbf{R}\mathbf{U} = \mathbf{R}\mathbf{U}\mathbf{H} = \mathbf{R}\mathbf{U}\mathbf{R}^T\mathbf{R}\mathbf{H} \tag{7A.11}$$

if $\mathbf{H} \in \mathcal{g}_{\mathbf{N}}$. Since $\mathbf{H}$ is orthogonal, so are $\mathbf{H}\mathbf{R}$ and $\mathbf{R}\mathbf{H}$. By the uniqueness of the polar decomposition, (7A.11) is thus equivalent to the following two separate conditions:

$$\mathbf{H}\mathbf{R} = \mathbf{R}\mathbf{H} \qquad \text{if } \mathbf{H} \in \mathcal{g}_{\mathbf{N}}, \tag{7A.12}$$

and

$$\mathbf{H}\mathbf{U} = \mathbf{U}\mathbf{H} \qquad \text{if } \mathbf{H} \in \mathcal{g}_{\mathbf{N}}. \tag{7A.13}$$

The restrictions due to these conditions can be found by a standard procedure.

First, we consider the condition (7A.12). For this purpose, we introduce the subgroup $\mathcal{h}_{\mathbf{H}}$ of orthogonal tensors $\mathbf{R}$ that commute with the fixed orthogonal tensor $\mathbf{H}$:

$$\mathcal{h}_{\mathbf{H}} \equiv \{\mathbf{R}: \mathbf{R} \in \mathcal{o}, \mathbf{R}\mathbf{H} = \mathbf{H}\mathbf{R}\}. \tag{7A.14}$$

Then (7A.12) is equivalent to

$$\mathbf{R} \in \bigcap_{\mathbf{H}\in\mathcal{g}_{\mathbf{N}}} \mathcal{h}_{\mathbf{H}} = \mathcal{r}_{\mathbf{N}} \qquad \text{where } \mathcal{r}_{\mathbf{N}} \equiv \bigcap_{\mathbf{H}\in\mathcal{g}_{\mathbf{N}}} \mathcal{h}_{\mathbf{H}}. \tag{7A.15}$$

Clearly, $\imath_N$ is also a subgroup, and its elements are the tensors **R** that satisfy the condition (7A.12). Since $\hbar_H = \hbar_{-H}$ and $\hbar_H = -\hbar_H$, to determine the subgroup $\imath_N$ we need to consider proper orthogonal tensors only. The restriction of the orthogonal subgroup $\hbar_H$ to a proper orthogonal subgroup will be denoted by $\hbar_H{}^+$. It is easily seen that there are then three possibilities:

Case 1. **H** is a rotation of angle 0, *i.e.*, $\mathbf{H} = \mathbf{1}$. Then obviously

$$\hbar_{\mathbf{1}} = \mathcal{O}. \tag{7A.16}$$

Case 2. **H** is a rotation of angle π about some axis $\mathbf{e}_1$. Then there exists an orthonormal basis $\{\mathbf{e}_i,\ i = 1, 2, 3\}$ such that

$$[H_{ij}] = \operatorname{diag}(1, -1, -1). \tag{7A.17}$$

Then

$$\hbar_H{}^+ = \{\mathbf{R}_{\mathbf{e}_1}{}^{\phi}\} \cup \{\mathbf{R}_{\mathbf{e}}{}^{\pi},\ \mathbf{e} = \alpha\mathbf{e}_2 + \beta\mathbf{e}_3,\ \alpha^2 + \beta^2 = 1\}. \tag{7A.18}$$

Here the symbol $\mathbf{R}_{\mathbf{e}}{}^{\alpha}$ denotes a rotation about the vector **e** through the angle α, and ϕ is an arbitrary real number.

Case 3. **H** is a rotation of angle α which is different from 0 or π. If $\mathbf{H} = \mathbf{R}_{\mathbf{e}}{}^{\alpha}$, then

$$\hbar_H{}^+ = \{\mathbf{R}_{\mathbf{e}}{}^{\phi}\}, \tag{7A.19}$$

where ϕ is an arbitrary real number.

Having considered these three cases in general, we can now give subgroups $\imath_N$ for various particular symmetry types. It should be noted that if $\imath_N$ is a set of generators for $\mathcal{G}_N$, then

$$\imath_N = \bigcap_{\mathbf{H} \in \mathcal{G}_N} \hbar_H. \tag{7A.20}$$

Specifically, we consider the symmetry types corresponding to the thirty-two crystalline classes, identifying them by the reference numbers used by TRUESDELL & NOLL†, and we adjoin to their list the transversely isotropic type (Ref. No. 12) and the isotropic type (Ref. No. 13). The results are given in Table 1.

In this table, (**i**, **j**, **k**) denotes a fixed orthonormal basis,

$$\mathbf{p} \equiv \sqrt{\tfrac{1}{3}}\,(\mathbf{i} + \mathbf{j} + \mathbf{k}),$$

ϕ is an arbitrary angle of rotation, and α, β are real numbers such that $\alpha^2 + \beta^2 = 1$. Of course, the complete subgroup $\imath_N$ is given by

$$\imath_N = \imath_N{}^+ \cup (-\imath_N{}^+). \tag{7A.21}$$

† *Cf. The Non-Linear Field Theories of Mechanics*, *op. cit.*, p. 83, where the classical crystallographic terms may be found.

TABLE 1

Ref. no.	*Generators of* $\mathscr{g}_N^+$	$\imath_N^+$
1	$\mathbf{1}$	$\mathscr{o}$
2	$\mathbf{R}_k^{\pi}$	$\{\mathbf{R}_k^0\} \cup \{\mathbf{R}_e^{\pi}, \mathbf{e} = \alpha\mathbf{i} + \beta\mathbf{j}\}$
3	$\mathbf{R}_i^{\pi}, \mathbf{R}_j^{\pi}$	$\{\mathbf{1}, \mathbf{R}_k^{\pi}\}$
4	$\mathbf{R}_k^{\frac{1}{2}\pi}$	$\{\mathbf{R}_k^{\phi}\}$
5	$\mathbf{R}_k^{\frac{1}{2}\pi}, \mathbf{R}_i^{\pi}$	$\{\mathbf{1}, \mathbf{R}_k^{\pi}\}$
6	$\mathbf{R}_i^{\pi}, \mathbf{R}_j^{\pi}, \mathbf{R}_p^{\frac{2}{3}\pi}$	$\{\mathbf{1}\}$
7	$\mathbf{R}_i^{\frac{1}{2}\pi}, \mathbf{R}_j^{\frac{1}{2}\pi}, \mathbf{R}_k^{\frac{1}{2}\pi}$	$\{\mathbf{1}\}$
8	$\mathbf{R}_k^{\frac{2}{3}\pi}$	$\{\mathbf{R}_k^{\phi}\}$
9	$\mathbf{R}_i^{\pi}, \mathbf{R}_k^{\frac{2}{3}\pi}$	$\{\mathbf{1}, \mathbf{R}_k^{\pi}\}$
10	$\mathbf{R}_k^{\frac{1}{3}\pi}$	$\{\mathbf{R}_k^{\phi}\}$
11	$\mathbf{R}_i^{\pi}, \mathbf{R}_k^{\frac{1}{3}\pi}$	$\{\mathbf{1}, \mathbf{R}_k^{\pi}\}$
12	$\{\mathbf{R}_k^{\phi}\}$	$\{\mathbf{R}_k^{\phi}\}$
13	$\mathscr{o}^+$	$\{\mathbf{1}\}$

Remark. We must be careful here not to draw the wrong conclusion that $\mathbf{K}_0$ is symmetric if and only if $\imath_N = \{\mathbf{1}, -\mathbf{1}\}$, since the polar decomposition (7A.10) of a symmetric tensor that is not positive-definite does not generally have so trivial an orthogonal part $\mathbf{R}$ as $\mathbf{1}$ or $-\mathbf{1}$.

We can follow a similar procedure to determine the restrictions on $\mathbf{U}$ due to the condition (7A.13). Here we define the *subspace* $\mathfrak{Z}_H$ of symmetric tensors $\mathbf{U}$ that commute with the fixed orthogonal tensor $\mathbf{H}$:

$$\mathfrak{Z}_H \equiv \{\mathbf{U}: \mathbf{U} \text{ symmetric}, \mathbf{UH} = \mathbf{HU}\}. \tag{7A.22}$$

Then we put

$$\bigcap_{\mathbf{H} \in \mathscr{g}_N} \mathfrak{Z}_H \equiv \mathfrak{U}_N, \tag{7A.23}$$

which is again a subspace. Now the range of the tensors $\mathbf{U}$ satisfying the condition (7A.13) is the set of all *positive-definite* members of $\mathfrak{U}_N$.

To determine the subspace $\mathfrak{U}_N$, we need the following:

Lemma. A symmetric tensor $\mathbf{U}$ commutes with an orthogonal tensor $\mathbf{H}$ if and only if $\mathbf{H}$ preserves the characteristic spaces of $\mathbf{U}$.

COLEMAN & NOLL have applied this lemma to determine the subspaces $\mathfrak{U}_N$ for the thirteen symmetry types to which Table 1 refers. We give their results in Table 2†.

Going back to (7A.10), we obtain the final results on the symmetry of $\mathbf{K}_0$ for a given type by multiplying all possible tensors $\mathbf{R}$ and $\mathbf{U}$ lying in the ranges given for that type in Tables 1 and 2. We summarize the results in Table 3.

Here the components K_{ij} are taken relative to the orthonormal basis

† *Cf. The Non-Linear Field Theories of Mechanics, op. cit.*, p. 85.

TABLE 2

Ref. no.	$\mathfrak{U}_N$
1	All symmetric tensors
2	$\mathbf{k}$ is a proper vector of $\mathbf{U}$
3	$\{a\mathbf{i} \otimes \mathbf{i} + b\mathbf{j} \otimes \mathbf{j} + c\mathbf{k} \otimes \mathbf{k}\}$
4	$\{a\mathbf{1} + b\mathbf{k} \otimes \mathbf{k}\}$
5	$\{a\mathbf{1}\}$
6	$\{a\mathbf{1}\}$
7	$\{a\mathbf{1}\}$
8	$\{a\mathbf{1} + b\mathbf{k} \otimes \mathbf{k}\}$
9	$\{a\mathbf{1} + b\mathbf{k} \otimes \mathbf{k}\}$
10	$\{a\mathbf{1} + b\mathbf{k} \otimes \mathbf{k}\}$
11	$\{a\mathbf{1} + b\mathbf{k} \otimes \mathbf{k}\}$
12	$\{a\mathbf{1} + b\mathbf{k} \otimes \mathbf{k}\}$
13	$\{a\mathbf{1}\}$

($\mathbf{i}$, $\mathbf{j}$, $\mathbf{k}$) in terms of which the symmetry group is specified, and a, b, c, d, e are arbitrary real numbers.

TABLE 3

Ref. no.	*Symmetry of* $\mathbf{K}_0$
1	$\mathbf{K}_0$ is arbitrary
2	$\|K_{ij}\| = \begin{Vmatrix} a & b & 0 \\ c & d & 0 \\ 0 & 0 & e \end{Vmatrix}$
4, 8, 10, 12	$\|K_{ij}\| = \begin{Vmatrix} a & b & 0 \\ -b & a & 0 \\ 0 & 0 & c \end{Vmatrix}$
3, 5, 6, 7, 9, 11, 13	$\mathbf{K}_0$ is necessarily symmetric

These results can be expressed as follows in terms of the classical crystallographic names:

1. Only in the triclinic system may the skew part of $\mathbf{K}_0$ be arbitrary.
2. With respect to suitable axes, two of the three possible independent components of the skew part of $\mathbf{K}_0$ must vanish in the monoclinic system, in three classes of the tetragonal system, in five of the hexagonal systems, and in transversely isotropic materials.
3. In the rhombic system, four classes of the tetragonal system, the cubic system, seven classes of the hexagonal system, and in isotropic materials, $\mathbf{K}_0$ is symmetric.

Remarks. A. The final results can be obtained also by direct calculations based on (7A.9), without assuming that $\mathbf{K}_0$ is non-singular.

B. By Fourier's inequality (7.8), the symmetric part of the tensor $\mathbf{K}_0$ must be positive-definite (or positive-semidefinite if it is not required to be non-singular). Thus there are additional restrictions on the constants $a, \ldots, e$ in the representations above, but these have no bearing on the skew part of $\mathbf{K}_0$.

Appendix 2

PHENOMENOLOGICAL ONSAGERISM IN PRACTICE

To cite all examples of the practice of choosing "fluxes" arbitrarily from the bilinear form giving the entropy production, would require several pages. I content myself with seven key examples.

1. The first general exposition of Onsagerism, by S. R. De Groot, *Thermodynamics of Irreversible Processes,* North Holland, Amsterdam, 1952. On page 7 De Groot gives the rule (7.42) as the "proper choice" of fluxes, but his immediately following summary makes no reference to it and gives the reader every encouragement to disregard it: " . . . the expression $\Delta\dot{S}$ can be split in several ways into a sum of conjugated J_i and X_i. But for all choices made, the Onsager relations (2) are valid among the coefficients L_{ik} of (1). Summarizing, we may say that the thermodynamic theory of an irreversible process of arbitrary complexity consists in finding first the conjugated fluxes and forces J_i and X_i from (6) [*viz* $\Delta\dot{S} = \Sigma J_i X_i$] by calculating the entropy production $\Delta\dot{S}$ and secondly in the study of the phenomenological equations (1) and the Onsager reciprocal relations (2)." This prescription is repeated on page 9. On page 8, "But the reader interested in the applications only can accept Onsager's theorem as a principle and continue with chapter III–XI," where, in fact, neither the rule (7.42) nor any other for "proper choice" is used for field variables (*e.g.*, page 98, "obviously the sum of the products of fluxes . . .").
2. The paper in which S. R. De Groot shows hydrodynamicists how "The viewpoint of the thermodynamics of irreversible processes . . . allows an insight into the fundamental concepts and the logical structure of hydrodynamics," *Proc. Symposia Appl. Math.*, **4**, *Fluid Dynamics,* pp. 87–99 (Am. Math. Soc.), McGraw-Hill, New York, 1953. This paper does not mention the rule (7.42) or any other "proper choice".
3. The treatise presenting, among other things, Onsagerism in detailed applications to chemistry: J. O. Hirschfelder, C. F. Curtiss, & R. B. Bird, *Molecular Theory of Gases and Liquids,* Wiley, New York, 1954. See Chapter 11, where the "proper choice" of forces and fluxes is described on page 706, although

it is not mentioned and does not apply in the "application" summarized by Table II.2-1 on pages 709 ff.

4. The definitive summary of experiments confirming the Onsager relations by D. MILLER, "Thermodynamics of irreversible processes, the experimental verification of the Onsager reciprocal relations," *Chem. Rev.*, **60**:15–17 (1960). This paper does not mention the rule (7.42) or any "proper choice".
5. Lest the above be dismissed as mere expositions, I cite also a research by the founder: L. ONSAGER, "Theories and problems of liquid diffusion," *Ann. N.Y. Acad. Sci.*, **46**:241–265 (1945); see pages 243–244, where fluxes are chosen with no mention of any test for eligibility.
6. Next an application by a leading expert on statistical mechanics: J. G. KIRKWOOD & B. CRAWFORD, JR., "The macroscopic equations of transport," *J. Phys. Chem.*, **56**:1048–1051 (1952); see page 1050: "The entropy source term . . . we shall arrange . . . to be in the form $\Phi = \Sigma \mathbf{j}_i \mathbf{F}_i$, where the $\mathbf{j}_i$ are 'currents' and the $\mathbf{F}_i$ are 'conjugate forces', gradients being very satisfactory as forces. This will permit us to apply Onsager's relations to the $\mathbf{j}_i$ and $\mathbf{F}_i$ so chosen."
7. Finally, an exposition by another of the founders: "Thermodynamik der irreversiblen Prozesse," pp. 413–523 of volume III/2 of FLÜGGE's *Handbuch der Physik*, Springer-Verlag, Berlin-Göttingen-Heidelberg, 1959. Here MEIXNER & REIK explain the Onsager-Casimir reciprocity relations on pages 425–426 for "scalar fluxes and the Cartesian components of vectorial or tensorial fluxes". The term "flux" first occurs on the preceding page, where it is said to include the heating-flux vector, the diffusion velocities, the pressure deviator, and the supplies of mass but is not otherwise defined or explained. The reciprocity relations are then stated, with no hint that the summands in the entropy production are to be split in anything but an arbitrary way. Moreover, nowhere in the article can I find any rule of selection.

Of these seven sources, only the last mentions "even" and "odd" variables, distinction between which seems to have become a *sine qua non* of more recent Onsagerism.

LECTURE 8

THE MAXWELL-BOLTZMANN EQUATION AS A CONSTITUTIVE EQUATION OF CONTINUUM MECHANICS

While speculations that gross matter is a vast assembly of tiny grains of a few distinct kinds are of great age, the first mathematical theories sufficient to yield quantities rather than pictures are those of NAVIER, POISSON, and CAUCHY, in which fluids and solids are represented as static or nearly static arrays of point centers of force. Although correct if somewhat special equations were drawn from such hypotheses by those mathematicians, physicists later rejected the molecular models on which they rest, entirely in the case of fluids, and with such qualifications for solids as to bury the starting point. Scientists should use these theories as examples to show that one man's molecules may be another man's poison.

The kinetic theory of gases, based on a different molecular model and a different concept of observation, was put into definitive form by MAXWELL and is the oldest granular theory still current today. It is designed to represent a moderately rarefied, monatomic gas. Apart from the molecular mass m and such other parameters as may serve to define a unique outcome to an encounter between two molecules (*e.g.*, a molecular diameter or the constants defining a law of intermolecular force), all observable quantities are stochastic expectations. These expectations are calculated from a probability assumed to be, at each place $\mathbf{x}$ and time t in four-dimensional Euclidean space-time, an absolutely continuous function of volume in a three-dimensional Euclidean space of vectors $\mathbf{v}$ regarded as molecular velocities. Thus the *number density* $n(\mathbf{x}, t)$, the expected number of molecules per unit volume at $\mathbf{x}$ and t, is given by

$$n = \int F(\mathbf{x}, \mathbf{v}, t)\, d\mathbf{v}, \tag{8.1}$$

where the integration extends over the entire three-dimensional space of molecular velocities $\mathbf{v}$, and the density $F(\mathbf{x}, \mathbf{v}, t)$ is called the *molecular density function*. Expectations based upon the probability defined by F we shall denote by an overbar:

$$\begin{aligned} \bar{f} \equiv \bar{f}(\mathbf{x}, t) &\equiv \frac{1}{n(\mathbf{x}, t)} \int f(\mathbf{x}, \mathbf{v}, t) F(\mathbf{x}, \mathbf{v}, t)\, d\mathbf{v}, \\ &= \frac{1}{n} \int fF\, d\mathbf{v}, \end{aligned} \tag{8.2}$$

where the second, shorter notation will be used except when it might give rise to confusion. (Of course, $F \geqq 0$, so $n \geqq 0$, and we assume here that

$n > 0$.) These expectations are to be used like the fields of continuum mechanics. While conceptually they stand for quantities in some unspecified sense "likely" to be observed, it seems that in physical theories the difference between "likely" and "certain" is only in the words. In particular, the kinetic theory affords statements about these expectations alone; if we do not find in these expectations what we wish, the kinetic theory does not show us where else to turn.

Fields intended to correspond with those common in continuum mechanics are defined as particular expectations:

$$\begin{aligned} \text{Mass density } \rho &\equiv mn, \\ \text{Stream velocity } \dot{\mathbf{x}} &\equiv \bar{\mathbf{v}}, \\ \text{Relative velocity } \mathbf{c} &\equiv \mathbf{v} - \dot{\mathbf{x}}, \\ \text{Pressure tensor } P &\equiv \overline{\rho\mathbf{c} \otimes \mathbf{c}}, \\ \text{Energetic } \varepsilon &\equiv \tfrac{1}{2}\overline{c^2}, \\ \text{Heating-flux vector } \mathbf{q} &\equiv \rho \cdot \tfrac{1}{2}\overline{c^2\mathbf{c}}. \end{aligned} \tag{8.3}$$

To apply these definitions in the context of Lecture 2, we are to interpret the pressure tensor $\mathbf{P}$ as the negative of the stress tensor $\mathbf{T}$, and the vector $\mathbf{q}$ as the negative of the heating-flux vector $\mathbf{h}$. Thus stress is nothing more than a gross manifestation of transfer of molecular momentum, relative to the stream velocity $\dot{\mathbf{x}}$, and likewise heating flux is nothing more than a gross manifestation of transfer of molecular kinetic energy. The energetic ε is the expected kinetic energy of the relative motion, per unit volume.

None of the quantities entering the kinetic theory bears a unit of temperature. In order to introduce the temperature θ, we must multiply it by a factor k such as to cancel the unit of temperature, so that $k\theta$ bears mechanical units only. Temperature, like energetic, is regarded as but another name for expected kinetic energy of relative motion; that is, the kinetic gas is assumed to be "ideal" in all conditions and to have a constant specific heat c_v at constant specific volume:

$$\varepsilon = c_v\theta = \frac{c_v}{k} \cdot k\theta. \tag{8.4}$$

The "empirical" rule that the specific heats of ideal gases vary inversely as the molecular masses suggests that†

$$\frac{c_v}{k} = \frac{3}{2} \cdot \frac{1}{m}, \tag{8.5}$$

† The relation (8.5) is set down, not proved, since in (8.4) the dimensions of k in mass, length, and time are not yet fixed.

where of course the factor $\frac{3}{2}$ expresses merely a conventional choice of units, and where k is the same for all gases. From (8.5), (8.4), and $(8.3)_5$ we see that

$$k\theta = \tfrac{1}{3}m\overline{c^2}. \tag{8.6}$$

(This formula often is taken as the definition of $k\theta$, and then (8.4) and (8.5) follow.) Moreover, if we let p denote the *mean pressure*,

$$p \equiv \tfrac{1}{3}\operatorname{tr}\mathbf{P}, \tag{8.7}$$

then from (8.6) and $(8.3)_4$ it follows that *the "ideal gas law" holds for every condition of the gas:*

$$p = kn\theta = \frac{k}{m}\rho\theta = \tfrac{2}{3}\rho\varepsilon. \tag{8.8}$$

This is not all. The pressure vector acting upon a surface with unit normal $\mathbf{n}$ is $\mathbf{Pn}$, and by $(8.3)_4$

$$\mathbf{n}\cdot\mathbf{Pn} = \rho\overline{(\mathbf{c}\cdot\mathbf{n})^2} \geqq 0. \tag{8.9}$$

Thus no part of any surface, be it a real boundary surface or a merely imagined one within the gas, can suffer tension in any flow.

Of course, one of the aims of the kinetic theory is to explain the viscosity of gases. Such an explanation presumes that by some process we can derive, in some sense of approximation, the Navier-Poisson-Stokes formulae (2.31) and (2.33), which in the present notation may be written as

$$\mathbf{P} = \varpi\mathbf{1} - \lambda(\operatorname{tr}\mathbf{D})\mathbf{1} - 2\mu\mathbf{D}, \tag{8.10}$$

where $2\mathbf{D} = \operatorname{grad}\dot{\mathbf{x}} + (\operatorname{grad}\dot{\mathbf{x}})^T$, and where ϖ, λ, and μ are functions of ρ and θ. The quantity ϖ is the *equilibrium pressure*, that is, that pressure which would equilibrate the gas, were it at rest at the density ρ and the temperature θ. Taking the trace of (8.10) yields by (8.8)

$$\frac{k}{m}\rho\theta - \varpi = -(\lambda + \tfrac{2}{3}\mu)\operatorname{tr}\mathbf{D}. \tag{8.11}$$

If $\lambda + \frac{2}{3}\mu \neq 0$, (8.11) implies the ridiculous conclusion

$$\operatorname{tr}\mathbf{D} = f(\rho, \theta); \tag{8.12}$$

that is, the expansion is determined by the density and temperature. In order for arbitrary expansions at any given density and temperature to be possible, it is thus necessary that the *Stokes relation* hold:

$$\lambda + \tfrac{2}{3}\mu = 0. \tag{8.13}$$

Consequently, by (8.11),

$$\varpi(\rho, \theta) = p = \frac{k}{m}\rho\theta: \tag{8.14}$$

The kinetic gas, to the extent it may be approximated as a viscous fluid, obeys the Stokes relation, and, in any flow, the mean pressure at any place and any time is the equilibrium pressure corresponding to the density and temperature at that place and time†.

Everything said so far follows from the definitions, before a word is said about the nature of the probability assigned to various molecular conditions or its evolution in time.

The density of molecules with given velocity $\mathbf{v}$ and position $\mathbf{x}$ at time t changes in two ways: by the motions the molecules undergo from the action of a field of body force $\mathbf{b}$ per unit mass, and by the motions resulting from forces the molecules exert on each other. The rates of increase of these two kinds equal one another, since molecules are not created or destroyed:

$$\partial_t F + \mathbf{v} \cdot \partial_{\mathbf{x}} F + \mathbf{b} \cdot \partial_{\mathbf{v}} F = \mathbb{C}[F], \tag{8.15}$$

where $\mathbb{C}[F]$ is the rate of increase of F due to molecular interactions. In the kinetic theory, these interactions are called "collisions" and are visualized as motions of pairs of molecules subject to their mutual attraction or repulsion alone. On the one hand, the gas is regarded as a vast multitude of speeding molecules; on the other, to describe the details of a collision, the entire universe is supposed empty except for two molecules. In virtue of this drastic simplification of mechanics as well as of daring stochastic assumptions, MAXWELL inferred a particular form for $\mathbb{C}[F]$. Write

$$\begin{aligned} F' &\equiv F(\mathbf{x}, \mathbf{v}', t), \\ F^* &\equiv F(\mathbf{x}, \mathbf{v}^*, t), \\ F^{*\prime} &\equiv F(\mathbf{x}, \mathbf{v}^{*\prime}, t), \\ \mathbf{w} &\equiv \mathbf{v}^* - \mathbf{v}. \end{aligned} \tag{8.16}$$

† It is easy to show that (8.10) cannot hold in the kinetic theory except in some kind of limit in which $\mathbf{D}$ (and presumably other quantities as well) becomes small. Indeed, by (8.13) and (8.9) we conclude from (8.10) that if $\mu > 0$,

$$\frac{(k/m)\rho\theta}{2\mu(\rho, \theta)} \geqq \mathbf{n} \cdot \mathbf{D}\mathbf{n} - \tfrac{1}{3} \operatorname{tr} \mathbf{D}$$

for all unit vectors $\mathbf{n}$ and all symmetric tensors $\mathbf{D}$. This inequality asserts that at a given density and temperature, certain velocity gradients are impossible, which is manifestly absurd, since nothing in the kinetic theory prevents us from choosing $F(\mathbf{x}, \mathbf{v}, 0)$ in such a way that (8.3) yields any given velocity field $\dot{\mathbf{x}}(\mathbf{x}, 0)$ at $t = 0$.

To get (8.13) from (8.11), it suffices to consider only $\mathbf{D}$ of arbitrarily small magnitude. Thus none of the "physical" evasions about "small" quantities can serve to squirm out of the conclusions here drawn.

Then MAXWELL's result is

$$\mathbb{C}[F] = \int\int w(F'F^{*\prime} - FF^*)\, dS\, d\mathbf{v}^*, \qquad (8.17)$$

where $\int \cdots dS$ stands for integration with respect to surface area over the part of the plane normal to the relative velocity $\mathbf{w}$ that corresponds to possible collisions, and where $\mathbf{v}'$ and $\mathbf{v}^{*\prime}$ are the asymptotic velocities at time $t = \infty$ of that pair of molecules which at time $t = -\infty$ had the asymptotic velocities $\mathbf{v}$ and $\mathbf{v}^*$, it being assumed here, as already stated, that the universe contains nothing but these two molecules. If a law of intermolecular force such as to deliver a unique pair $\mathbf{v}'$, $\mathbf{v}^{*\prime}$ corresponding to an arbitrary pair $\mathbf{v}$, $\mathbf{v}^*$ be laid down, then (8.17) defines a functional operation, provided the integrals converge. This last requirement can be relaxed, since F itself is never needed in the kinetic theory, which concerns only expectations based on the probability defined by F. Thus it is only an integrated form of (8.15), as given by MAXWELL, that is important, but (8.15) itself, which was derived by BOLTZMANN, is formally more perspicuous; with $\mathbb{C}[F]$ as given by (8.17), it is called the *Maxwell-Boltzmann equation.*

As the theory presently stands, (8.17) is a *definition,* in the same sense that the formula $\mathbf{P} = p\mathbf{1}$ is a definition in Eulerian hydrodynamics. Both formulae can be and have been motivated in various ways—the arguments of MAXWELL and BOLTZMANN cannot be assessed rationally as anything but heuristic. EULER's definition has been proved to agree with the first approximation to a general class of constitutive relations in continuum mechanics, according to a rigorous process of approximation. While attempts to prove a corresponding theorem for MAXWELL's definition (8.17) have not yet been successful,† we have every reason to expect that in time it, too, will be given a rational status.

In these lectures I shall discuss, not the foundations of the kinetic theory and the origins and significance of its basic law, but rather some of the simplest consequences and applications of that law. To set a proper value upon such consequences and applications, it is necessary to understand first how special that law is. While the founders believed, or at least hoped, that they were discovering the physics governing molecular motions, we know today they were not. Physicists no longer regard the molecules of even helium or hydrogen as mathematical points or spheres following the laws of the Newtonian mechanics of conservative systems. Moreover, the kinetic theory contradicts those laws, for they obey the reversibility theorem, while the kinetic theory, as we shall see in a moment, represents the gas as a dissipative material, not just sometimes or proba-

† I refer, of course, to attempts to "derive" the Maxwell-Boltzmann equation by some limiting process from what specialists in statistical mechanics rather confusingly call "Liouville's equation".

bly, but invariably and certainly. Furthermore, as is plain from (8.17), the kinetic theory rests upon a peculiar stochastic hypothesis of quasi-equilibrium: The pairs are statistically independent not only before but also after each encounter. Such independence contradicts the laws of ordinary mass-point mechanics on the molecular scale, since those laws demand specific outcomes determined by initial conditions. We may describe the Maxwell-Boltzmann collision axiom as expressing a special kind of fading memory which makes the pairs of molecules instantly lose the correlations their encounters give them.

While the kinetic theory violates Newtonian molecular mechanics, it is consistent with classical continuum mechanics and affords a special case of it. Here lies the peculiar interest of the kinetic theory, the fascination it exerts upon every thoughtful student of mechanics. It is not a theory *of* molecules; rather, it affords us a means of taking a certain, very limited account of molecular constitution in formulating a continuum theory of gases. It is a theory whose outcome, ideally, is a constitutive equation of continuum mechanics, just one single constitutive equation, but one of such enormous complexity that we do not even try to find it except under special conditions. In this sense we may describe the kinetic theory as a theory of approximate constitutive equations, approximate in specified senses and specified circumstances. While most of the older efforts in continuum mechanics concerned solution of special cases within theories based on well established or at least widely accepted particular constitutive equations gotten from outside the theory itself, the kinetic theory rarely if ever takes up boundary-value or initial-value problems, its main if not whole concern being to find the constitutive equations. Now that modern continuum mechanics also has developed so far as to include theories of constitutive equations, rather than mere consequences of three or four special ones, the kinetic theory no longer seems so different as once it did. Modern continuum mechanics and the kinetic theory of monatomic gases seek the same end, namely, correct relations among gross fields such as the pressure tensor and the history of the temperature and velocity.

When it was thought that the kinetic theory described gases composed of real molecules, some pretty raw mathematics was allowed. Now that we see clearly the special nature of the kinetic theory, now that we understand its function as being a particular, highly simplified mathematical model for irreversibility in gas flows, we must stop guessing and get down to work on rigorous mathematical proofs, or at least on clear and systematic procedures on which we may hope someday to base rigorous proofs. The interest of the kinetic theory is purely rational, not physical; while "intuitive" or "physical" thought may be helpful in physics, botch-work in a rational theory destroys its rationality and hence its function.

In this lecture and the next two I will present some aspects of the kinetic theory as a branch of continuum mechanics. While the results sought are much the same in kind as those in ordinary continuum mechanics, they are entirely different in detail. First, different materials (*i.e.*, different gases) are specified by molecular masses and intermolecular force laws, not by gross properties, and the corresponding gross properties have to be extracted, somehow, from the theory of the Maxwell-Boltzmann equation. Second, moduli such as viscosity, which in ordinary continuum mechanics are regarded as disposable, in the kinetic theory have specific, unalterable forms and values, which we must find by mathematical proof and calculation. It is this concreteness, bought at the price of the extremely special molecular structure and stochastic hypotheses laid down from the start, that gives the kinetic theory its own charm.

As we have seen, the collision operator $\mathbb{C}[\quad]$ transforms a function of molecular velocity $\mathbf{v}$ into another one. The *total collision operator* $\mathbb{C}$ applied to a function $f(\mathbf{x}, \mathbf{v}, t)$ yields the expected change in that function due to all possible collisions when the molecular density before these collisions is F:

$$\mathbb{C}f \equiv \int f\mathbb{C}[F]\, d\mathbf{v}. \tag{8.18}$$

MAXWELL noticed that for certain functions f, it follows from (8.18) and (8.17) that

$$\mathbb{C}f = 0. \tag{8.19}$$

Such f are

$$f = 1, \mathbf{v}, \tfrac{1}{2}v^2. \tag{8.20}$$

That $\mathbb{C}f$ vanishes for these f follows from the requirement that mass, linear momentum, and kinetic energy be conserved in any law of encounter admitted in the kinetic theory†. Thus the vanishing of $\mathbb{C}f$ for these f should reflect the balance of mass, linear momentum, and total energy in the gas as a whole, no matter what solution F of the Maxwell-Boltzmann equation be considered. MAXWELL showed that this is so.

Indeed, if we multiply the Maxwell-Boltzmann equation (8.15) by $f(\mathbf{v})$ and integrate over $\mathbf{v}$-space, we obtain MAXWELL's *equation of transfer:*

$$(\rho\bar{f})^{\cdot} + \rho\bar{f}\,\operatorname{div}\dot{\mathbf{x}} + \operatorname{div}(\rho\overline{\mathbf{c}f}) = -m\int f\mathbf{b}\cdot\partial_{\mathbf{v}}F\, d\mathbf{v} + m\mathbb{C}f, \tag{8.21}$$

where the superior dot on the first summand denotes the substantial derivative, the time derivative following the stream velocity $\dot{\mathbf{x}}$.

If in this equation we put $f = 1, \mathbf{v}$, and $\frac{1}{2}v^2$ in turn and use (8.19), then by expressing the results in terms of the definitions (8.3), (8.4), and (8.6),

† In virtue of BOLTZMANN's lemma, stated and proved below, $\mathbb{C}f$ vanishes *only* for linear combinations of these particular f.

we obtain the following *field equations*†:

$$\begin{aligned} \dot{\rho} + \rho \operatorname{div} \dot{\mathbf{x}} &= 0, \\ \rho\ddot{\mathbf{x}} &= -\operatorname{div} \mathbf{P} + \rho\mathbf{b}, \\ \rho\dot{\varepsilon} &= -\operatorname{tr}(\mathbf{P} \operatorname{grad} \dot{\mathbf{x}}) - \operatorname{div} \mathbf{q}. \end{aligned} \tag{8.22}$$

These field equations are of just the same form as those expressing the balance of mass, linear momentum, and total energy in the continuum mechanics of a single medium devoid of body heating, namely (2.3), (2.8), and (2.18). Moreover, the definition (8.3) shows that $\mathbf{P}$ is symmetric, in accord with (2.10). Thus the kinetic gas is in all respects a special kind of continuum as far as *pure mechanics* is concerned.

Thermodynamics is another matter. Entropy has not yet been defined, so we cannot even ask about caloric equations of state. To arrive at a basis for thermodynamics, we first need a concept of equilibrium.

MAXWELL regarded as being appropriate to equilibrium any molecular-density function F such as to be left unaltered by collisions:

$$\mathbb{C}[F] = 0. \tag{8.23}$$

He inferred, and others have since proved rigorously, that any such F is of the form

$$F = F_{\mathrm{M}} \equiv ae^{-bc^2}, \tag{8.24}$$

where $\mathbf{c} \equiv \mathbf{v} - \dot{\mathbf{x}}$, $\dot{\mathbf{x}}$ being an arbitrary function of $\mathbf{x}$ and t, while a and b are arbitrary positive functions of $\mathbf{x}$ and t. Of course, $\dot{\mathbf{x}}$ is the stream velocity, and by (8.6) and $(8.3)_1$

$$b = \frac{m}{2k\theta}, \qquad a = \frac{\rho}{m(2\pi k\theta/m)^{\frac{3}{2}}}. \tag{8.25}$$

Such an F is called a *Maxwellian molecular density*. For it, the pressure is hydrostatic, and the heating flux vanishes:

$$\mathbf{P} = p\mathbf{1}, \qquad \mathbf{q} = \mathbf{0}, \tag{8.26}$$

as is plain from the definitions $(8.3)_{4,6}$ and (8.7).

In order to satisfy the Maxwell-Boltzmann equation (8.15), F_{M} must satisfy also the equation

$$\partial_t F_{\mathrm{M}} + \mathbf{v} \cdot \partial_{\mathbf{x}} F_{\mathrm{M}} + \mathbf{b} \cdot \partial_{\mathbf{v}} F_{\mathrm{M}} = 0. \tag{8.27}$$

† We assume further that $fbF = o(v^{-2})$ as $v \to \infty$. While of course such an assumption may be imposed at any one time, there is no guarantee that it will then hold at subsequent times. Thus MAXWELL's derivation of $(8.22)_2$ and $(8.22)_3$, though it has passed unquestioned to this day in the literature, must be regarded as lacunary except in the case when $\mathbf{b} = \mathbf{0}$.

In the special case when $\mathbf{b} = \mathbf{0}$, we easily prove that the most general Maxwellian solution of (8.27) corresponds to the velocity field

$$\dot{\mathbf{x}} = [\tfrac{1}{3}d(t)\mathbf{1} + \mathbf{W}(t)]\mathbf{p} + \mathbf{e}(t), \tag{8.28}$$

where the scalar $d(t)$ is the expansion, the skew tensor $\mathbf{W}(t)$ is the spin, and $\mathbf{p}$ is the position vector with respect to some origin. Such a flow represents a uniform dilatation superimposed upon a rigid motion. Since $\mathbb{C}[F_{\mathrm{M}}] = 0$, flows of this special kind are dissipationless. We may see this in another way by noticing that (8.26) reduces the energy equation $(8.22)_3$ to the form

$$\rho\dot{\varepsilon} = -pd. \tag{8.29}$$

There are now two possibilities. First, suppose $d(t) = 0$ in an interval of time. The motion is then rigid. By $(8.22)_1$, $\rho =$ const.; by (8.29), $\varepsilon =$ const.; by (8.8), $\theta =$ const. and $p =$ const. Thus in a rigid motion the kinetic gas as a whole behaves exactly like an Eulerian gas. If $d(t) \neq 0$ in an interval of time, by (8.8) we obtain from (8.29) a differential equation whose general solution is

$$p\rho^{-\frac{5}{3}} = \text{const.}, \tag{8.30}$$

so that these motions, interpreted in classical terms, are "adiabatic" expansions of an ideal perfect gas having 5:3 as the ratio of its specific heats, assumed constant. Thus these "locally Maxwellian" solutions are the trivial solutions in the sense that they correspond to the only gas flows for which the Maxwell-Boltzmann mechanism of irreversibility is neutralized. Until not long ago, they were the only exact solutions known in the kinetic theory.

Since these flows include arbitrary expansions† and rotations, they represent something a little broader than what we think of in ordinary mechanics as being equilibrium. WANG's definition of *local equilibrium* in the kinetic theory is

$$F(\mathbf{x}, \mathbf{v}, t) = F(\mathbf{x}, \mathbf{v}) = F(\mathbf{x}, -\mathbf{v}). \tag{8.31}$$

That is, F shall be steady and invariant under reversal of velocities. Extending results of MAXWELL and BOLTZMANN, WANG was able to prove‡ that if the body force is conservative with steady potential energy, say

† That uniform expansions are included in these "reversible" flows may be expected, both from the Stokes relation (8.13) and from heuristic considerations about ideal gases. In fact, $d(\cdot)$ and $\mathbf{e}(\cdot)$ must be linear functions, and $\mathbf{W}(\cdot)$ must be a constant function.

‡ The formula (8.32) goes back to the days of MAXWELL and BOLTZMANN, but neither they nor later writers defined what they meant by kinetic equilibrium in a field of force or proved that (8.32) is the only solution that corresponds to it.

$\phi(\mathbf{x})$, then in local equilibrium

$$F = ae^{-(2b/m)(\frac{1}{2}mv^2+\phi)} = F_{\mathrm{M}}e^{-(2b/m)\phi}, \tag{8.32}$$

where F_{M} is a Maxwellian density whose parameters a and b are positive constants and whose stream velocity $\dot{\mathbf{x}}$ is zero. By $(8.26)_1$ the field equation $(8.22)_2$ reduces to EULER's equation of hydrostatic equilibrium. Thus the kinetic theory yields for equilibrium, duly defined, exactly the same results as does the hydrostatics of a corresponding ideal gas.

The main tool used by WANG in his proof is BOLTZMANN's ***irreversibility lemma:*** For any positive function F,

$$\mathbb{C}\log F \begin{matrix} < 0 & \text{if } F \text{ is not Maxwellian,} \\ = 0 & \text{if } F \text{ is Maxwellian.} \end{matrix} \tag{8.33}$$

While some writers, particularly those not really conversant with the kinetic theory, represent this lemma as somehow proving that molecular collisions result in irreversibility, such is not at all the case, for the lemma concerns the functional operator $\mathbb{C}$ and does not require that F solve the Maxwell-Boltzmann equation. It reflects the extremely special nature of the collision operator $\mathbb{C}$ and shows that the *assumptions* of the kinetic theory regarding the effect of collisions on the probability build into it a particular irreversibility: No matter what be F, unless it be Maxwellian, the whole effect of collisions is to diminish $\log F$. Since the collision laws of the mechanics of mass-points satisfy the reversibility theorem, the basic assumption of the kinetic theory strictly excludes the mechanical model laid down as a heuristic basis. The most interesting problems in the kinetic theory refer to the effects of this special and somewhat mysterious mechanism of irreversibility.

The proof of BOLTZMANN's lemma is simple. First one derives the identity

$$\mathbb{C}\log F = \frac{1}{4}\iiint w\left(\log\frac{FF^*}{F'F^{*\prime}}\right)(F'F^{*\prime} - FF^*)\,dS\,d\mathbf{v}^*\,d\mathbf{v}, \tag{8.34}$$

which follows from the definition (8.17) by some easy transformations. The two factors of the integral in (8.34) are of opposite sign unless each vanishes. Thus the inequality follows. Conversely, if $\mathbb{C}\log F = 0$, it is necessary that each factor vanish. If so,

$$\log F + \log F^* = \log F' + \log F^{*\prime}. \tag{8.35}$$

Any continuous solution† of this functional equation must be a linear

† The theorem and a proof of it, for twice continuously differentiable functions, are due to BOLTZMANN; as stated above, to GRONWALL; an elegant proof is given by H. GRAD, Appendix 1 to "On the kinetic theory of rarefied gases," *Comm. Pure Appl. Math.* (*N.Y.U.*), **2**:231–407 (1949).

combination of the mass, momentum, and kinetic energy of the molecule. Hence F is Maxwellian.

BOLTZMANN's lemma (8.33) suggests that the expectation of $\log F$, when F is made to be a solution of the Maxwell-Boltzmann equation, will be a gross quantity subject to a definite trend in time. In classical thermodynamics, the calory ("entropy") is such a quantity. To connect the two ideas, we seek a proper definition for the caloric in the kinetic theory, using what we have just shown. That is, we identify kinetic equilibrium, and hence an appropriate Maxwellian distribution, with the thermostatic equilibrium of an ideal gas having 5:3 as its ratio of specific heats, assumed to be constant. In thermostatic equilibrium of such a gas we have the standard equation of state for the caloric η, which we shall now write as η_{M}:

$$\eta = \eta_{\mathrm{M}} \qquad \text{where } \eta_{\mathrm{M}} \equiv \frac{k}{m} \log \frac{\theta^{\frac{3}{2}}}{\rho} + \text{const.} \tag{8.36}$$

But if F is Maxwellian, we see from (8.24) and (8.25) that

$$\begin{aligned} \overline{\log F_{\mathrm{M}}} &= \overline{\log a} - \overline{bc^2}, \\ &= \log a + \text{const.}, \\ &= \log \frac{\rho}{\theta^{\frac{3}{2}}} + \text{const.} \end{aligned} \tag{8.37}$$

Combining (8.36) and (8.37) shows that if we are to define the caloric η_{K} from F in such a way that the equilibrium theory of the kinetic gas be consistent with the thermostatics of ideal gases, then it is necessary that when F is Maxwellian†, $\eta = \eta_{\mathrm{K}}$, where

$$m\eta_{\mathrm{K}} \equiv -k\,\overline{\log F} + \text{const.} \tag{8.38}$$

BOLTZMANN regarded η_{K}, defined by (8.38), as being the caloric associated with the solution F of the Maxwell-Boltzmann equation, *no matter what that solution was.* If we define the kinetic calory H in a finite region $\mathcal{V}$ as

$$H \equiv \int_{\mathcal{V}} \rho\eta_{\mathrm{K}}\, dV \tag{8.39}$$

and set

$$r \equiv -k\mathbb{C} \log F, \tag{8.40}$$

then from the Maxwell-Boltzmann equation (8.15) we easily conclude that

$$\dot{H} - \frac{k}{m} \int_{\partial\mathcal{V}} \overline{\mathbf{c} \log F} \cdot \mathbf{n}\, dA = \int_{\mathcal{V}} r\, dV, \tag{8.41}$$

† While works on irreversible thermodynamics usually give the particular value k to the constant, it is unnecessary to do so for any result derived. Attempts to make the "zero of entropy" a meaningful concept seem to be futile for any concrete use.

where, by BOLTZMANN's lemma (8.33), the right-hand side is positive unless F is Maxwellian. That is,

$$\begin{aligned} \dot{H} - \frac{k}{m}\int_{\partial \mathcal{V}} \overline{\mathbf{c}\log F}\cdot \mathbf{n}\,dA &> 0 \qquad \text{if } F \text{ is not Maxwellian,} \\ \dot{H} &= 0 \qquad \text{if } F \text{ is Maxwellian.} \end{aligned} \tag{8.42}$$

The dot denotes the time derivative for a material volume, defined by the stream velocity $\dot{\mathbf{x}}$. This is ***Boltzmann's H-theorem,*** the main statement of gross irreversibility in the kinetic theory of gases.

Some early students of the kinetic theory regarded BOLTZMANN's theorem as a proof that a great assembly of molecules would in time settle into a state of equilibrium, in which its entropy would be the greatest possible and would change no more. Such a statement corresponds to one of the several different principles called, loosely, the "second law of thermodynamics". This, most certainly, is *not* the content of BOLTZMANN's H-theorem. We expect that the kinetic theory should be capable of dealing with at least all situations described by the simpler phenomenological theories of gases. Among these are steady flows such as that induced by the steady sliding motion of one infinite plate relative to another, each being heated to a prescribed temperature. Such a flow may not correspond to a thermodynamicist's model of the "universe", but it is certainly convenient and useful in fluid mechanics. There is no "trend to equilibrium" here, nor should there be. BOLTZMANN's H-theorem follows from the fully general kinetic theory and thus, presumably, should apply to such a flow as this. Thus the H-theorem in its general form certainly ought not to imply a universal trend to equilibrium; presumably it does not.

Indeed, there are symmetry arguments such as to persuade us that $\overline{\mathbf{c}\log F}$ on the wall of a closed vessel is tangent to the wall, so that the surface integral vanishes, and the H-theorem then asserts that

$$\begin{aligned} &H \text{ increases if } F \text{ is not Maxwellian,} \\ &H \text{ is constant if } F \text{ is Maxwellian.} \end{aligned} \tag{8.43}$$

While we see a formal similarity to CLAUSIUS' claim that the entropy of the universe is increasing to a maximum, there are troubles. The statement of CLAUSIUS presumes an assigned, fixed total energy. The kinetic gas is not a system of fixed energy, as we may see from a glance at its equation of balance of energy $(8.22)_3$. Although, of course, a molecular encounter conserves energy, the collision operator is such that the total energy of the gas is not conserved, but rather is gained in the volume at the rate given by

$$\dot{\mathfrak{K}} + \dot{\mathfrak{E}} = -\int_{\partial\mathcal{V}} (\dot{\mathbf{x}} \cdot \mathbf{P}\mathbf{n} + \mathbf{q} \cdot \mathbf{n})\, dA = -\int_{\partial\mathcal{V}} \rho\overline{(\dot{\mathbf{x}} \cdot \mathbf{c} + \tfrac{1}{2}c^2)\mathbf{c}} \cdot \mathbf{n}\, dA, \tag{8.44}$$

when $\mathbf{b} = \mathbf{0}$. Indeed, the total energy $\mathfrak{K} + \mathfrak{E}$ is conserved for an *isolated system*, namely, one in which

$$\overline{(\dot{\mathbf{x}} \cdot \mathbf{c} + \tfrac{1}{2}c^2)\mathbf{c}} \cdot \mathbf{n} = 0 \qquad \text{on } \partial\mathcal{V}, \tag{8.45}$$

and this condition is satisfied if $\dot{\mathbf{x}} = \mathbf{0}$ and F is even in $\mathbf{c}$ on $\partial\mathcal{V}$, but the kinetic theory does not give us any reason to expect this condition to hold on any particular boundary or in the limit for an infinite region for any

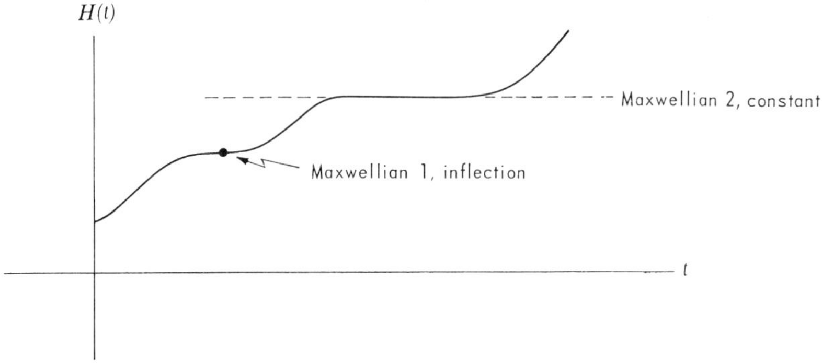

FIGURE 9. Sketch to illustrate behavior of H(·) compatible with the H-theorem.

particular solution except, of course, such as are even in $\mathbf{c}$ everywhere. Thus MAXWELL's kinetic theory seems dubious as a model for CLAUSIUS' universe.

In any case, in the kinetic theory the irreversibility is put in by the theorist, not proved by him, for, as we have seen, it is the *assumed form* of the collision operator $\mathbb{C}[\ \ ]$, not the temporal evolution of F according to the Maxwell-Boltzmann equation, that leads to BOLTZMANN's lemma, which is the H-theorem in little. Second, as we have seen also, a Maxwellian molecular density is only necessary, not sufficient for equilibrium in the kinetic theory. A Maxwellian solution may correspond to time-dependent expansions and rigid rotations, scarcely to be associated with equilibrium. Third, since there are infinitely many different Maxwellian solutions, the H-theorem (8.43) as it stands does not forbid an H-curve such as is shown in Figure 9. Fourth, even if H does approach a limit in time, and that limit corresponds to a Maxwellian F, it does not follow that

F itself approaches a Maxwellian density, and hence it does not follow that other quantities, such as the pressure tensor and the heating-flux vector, approach forms appropriate to equilibrium. The group of questions connected with the H-theorem in the special form (8.43) is called the *problem of the trend to equilibrium*. It is the subject of the next lecture.

If we return now to the general H-theorem (8.42), it is natural to set it alongside the Clausius-Duhem inequality (2.50), which in the present case and notation is

$$\dot{\mathfrak{H}} + \int_{\partial\mathcal{V}} \frac{\mathbf{q}\cdot\mathbf{n}}{\theta}\, dA \geqq 0. \tag{8.46}$$

In order for the two to agree for all regions, by $(8.3)_6$ and (8.6) it is necessary and sufficient that

$$\tfrac{3}{2}\rho\overline{c^2\mathbf{c}}/\overline{c^2} = -\rho\overline{\mathbf{c}\log F} + \text{a solenoidal field.} \tag{8.47}$$

Clearly only certain particular functions F have their expectations so related. The Maxwell-Boltzmann equation controls only the evolution of F in the course of time, not placing any restriction on F at some one fixed time. Since (8.47) is a time-independent relation not satisfied by all F, *a fortiori* it cannot be satisfied by all solutions of the Maxwell-Boltzmann equation. Indeed, for one particular class of solutions, the Maxwellian ones, (8.47) is satisfied, since

$$\tfrac{3}{2}\rho c^2\mathbf{c}/\overline{c^2} = -\rho\mathbf{c}\log F + \rho\mathbf{c}\log a, \tag{8.48}$$

but for this class (8.46) becomes trivial. These facts show that *the H-theorem and the Clausius-Duhem inequality are not the same*†, though they may be expected to fall into approximate consonance for gas flows sufficiently near to equilibrium in the sense that F differs little from the corresponding F_{M}.

This does not show that either principle is right or wrong. The Clausius-Duhem inequality, though of some age, has only recently been subjected to mathematical study in contexts of any generality. We are not in a position to affirm it as a general law of gross physics. The special nature of the gas model and the special form of the collision operator in the kinetic theory forbid us from taking that theory, either, as basic. Finally, it remains possible that *both* statements of irreversibility be

† This fact is the basis for I. MÜLLER's replacement of the Clausius-Duhem inequality (8.46) by the less specific requirement

$$\dot{\mathfrak{H}} + \int_{\partial\mathcal{V}} \boldsymbol{\phi}\cdot\mathbf{n}\, dA \geqq 0,$$

in which the caloric flux $\boldsymbol{\phi}$ is to join $\mathbf{T}$ and $\mathbf{h}$ as a quantity delivered by a constitutive equation. *Cf.* p. 58, above, and Lecture 5.

correct, for one of two reasons; first, we may be wrong to take (8.38) as the definition of caloric in states other than equilibrium, and second, since irreversibility is a trend, not balance, there may be more than one correct statement of it.

Looking back at the Maxwell-Boltzmann equation,

$$\partial_t F + \mathbf{v} \cdot \partial_{\mathbf{x}} F + \mathbf{b} \cdot \partial_{\mathbf{v}} F = \mathbb{C}[F], \tag{8.15}$$

we remark that it is of the first order in the time and suggests an initial-value problem:

$$F(\mathbf{x}, \mathbf{v}, t) = \mathbb{S}F(\mathbf{x}, \mathbf{v}, 0), \tag{8.49}$$

where $\mathbb{S}$ is the solution operator. If this is so, the molecular condition at time t will depend on that at time 0, as the molecular model itself suggests. However, only rarely does the kinetic theory face this problem, for it deals with molecules only to the extent that they influence gross specimens. Rather, it seeks to derive from the Maxwell-Boltzmann equation (8.15) time-independent relations among gross quantities: connections between the pressure tensor $\mathbf{P}$, the heating flux $\mathbf{q}$, the caloric η, and the fields of velocity, density (or pressure), and temperature. These relations are to be substituted into or at least used side by side with the field equations (8.22) expressing balance of mass, momentum, and energy. The Navier-Poisson-Stokes formula (8.10) is an example of such a relation; another is Fourier's law,

$$\mathbf{q} = -\kappa \operatorname{grad} \theta. \tag{8.50}$$

The earliest investigators forced these two constitutive equations of nineteenth-century continuum mechanics into agreement with the kinetic theory by casting away whatever terms in the equations of the latter were inconvenient. As necessary conditions for such compatibility resulted explicit formulae for the moduli λ, μ, and κ in terms of molecular quantities. These formulae are described as "evaluations" of viscosity and conductivity according to the kinetic theory. Later investigators introduced systematic methods whereby (8.10) and (8.50) appear as the first terms in expansions for $\mathbf{P}$ and $\mathbf{q}$; results of this kind are described by kinetic theorists as "corrections to hydrodynamics". Nowadays we should say rather that these procedures derive for the kinetic gas approximate or limiting constitutive equations of increasing refinement.

While in this lecture I cannot present these methods, elaborate and intricate as they are, I shall make a few remarks on their general quality.

The results sought, time-independent relations among gross quantities, can scarcely be expected to have more than summary status. The theory shows us that the way we start the molecular distribution influences it forever; hence different methods of starting the gas flow can lead to different gross quantities, and we cannot expect relations among them to be time-independent, except those introduced at the start by time-independent definitions. If, however, some particular solutions are singled out as being limits or asymptotic forms of infinite families of others, such special relations as hold for these singular solutions may have singular or common status for whole classes of solutions, even for all solutions. The integration procedures in the kinetic theory assume that such solutions, which may be called *dominant solutions*†, exist and seek them, and them alone. The procedures themselves, while systematic, are merely formal. The best known are those of HILBERT, CHAPMAN, and ENSKOG. While these lead to different approximations at the nth stage, GRAD showed in his famous article‡ that in the limit as $n \to \infty$ the results are the same for all. The proof is one of intermediate rigor: While expansions are assumed to converge and partial differential systems are assumed to possess unique solutions to initial-value problems, the calculations are formally exact, no terms being thrown away from caprice or "physical intuition". Assertions backed by proofs of this type, since they have a long history of being found later to be right, if subject, perhaps, to some conditions unsuspected at the outset, deserve to be looked upon with great respect.

The whole problem of the existence and status of dominant solutions in general remains, more than a century after the first groping toward such solutions by MAXWELL and BOLTZMANN themselves, covered by mystery and approached through a delicate combination of guesswork and staggering formal calculations. In these lectures, I cannot deal with it at all. I wish here merely to remark on the status of dominant solutions for the kinetic gas in terms of the general theory of continuous media today.

Traditionally, the dominant solutions are so arranged as to fix attention on the first five moments of F. These are

$$\rho, \qquad \text{the three components of } \rho\dot{\mathbf{x}}, \qquad 2\rho(\varepsilon + \tfrac{1}{2}\dot{\mathbf{x}}^2).$$

Anything that can be expressed in terms of them can be expressed equally in terms of

$$\rho, \qquad \theta, \qquad \dot{\mathbf{x}}.$$

Thus we may say that the dominant solutions seek to determine gross quantities such as $\mathbf{P}$ and $\mathbf{q}$ in terms of the fields of density, velocity, and temperature. The essence of the search for dominant solutions is to eliminate the time by means of the equation of transport (8.21), which gives explicitly the time derivative of any gross quantity, at a given instant, as

† Advisedly I do not use the term "normal solution", since writers on the kinetic theory associate it with particular procedures or particular functional forms.

‡ H. GRAD, "Principles of the kinetic theory of gases," in FLÜGGE's *Handbuch der Physik*, vol. XII, Springer-Verlag, Berlin-Göttingen-Heidelberg, pp. 205–294 (1958).

a function of certain space derivatives of appropriate gross quantities. Whether this is done explicitly, as in the Chapman-Enskog method, or implicitly by integrating a system of partial differential equations, as in the Hilbert method, a molecular-density function F is sought such that

$$F(\mathbf{x}, \mathbf{v}, t) = \mathfrak{f}_t(\rho(\mathbf{y}, 0), \theta(\mathbf{y}, 0), \dot{\mathbf{x}}(\mathbf{y}, 0); \mathbf{v}), \tag{8.51}$$

where $\mathfrak{f}_t$ denotes a time-dependent functional, where $\mathbf{y}$ varies over the field of flow†, and where $\mathbf{v}$ is a parameter. The nature of the functional $\mathfrak{f}_t$ is believed or at least hoped to be determined, in principle, by the Maxwell-Boltzmann equation. Thus (8.51) is a special case of a general memory relation‡:

$$F(\mathbf{x}, \mathbf{v}, t) = \mathfrak{f}(\rho^t, \theta^t, \dot{\mathbf{x}}^t; \mathbf{v}), \tag{8.52}$$

where the superscript t denotes a history, and where the history refers to a tube of trajectories of the stream velocity $\dot{\mathbf{x}}$, *i.e.*, to a neighborhood of a fixed particle in the sense of continuum mechanics. Since the gross quantities $\mathbf{P}$, $\mathbf{q}$, ε, and η are determined from F by quadrature according to $(8.3)_{4,5,6}$ and (8.38),

$$\{\mathbf{P}, \mathbf{q}, \varepsilon, \eta\} = \mathbb{F}(\rho^t, \theta^t, \dot{\mathbf{x}}^t), \tag{8.53}$$

where $\mathbb{F}$ denotes a functional. This relation is a special case of the *principle of determinism* of modern continuum mechanics. It should not be difficult to prove that (8.53), as derived from (8.51), satisfies also the principles of material frame-indifference and local action. If so, in view of the field equations (8.22) we see that *all* the requirements of modern continuum mechanics are satisfied by the dominant solutions in the kinetic theory (if they really exist), *except the Clausius-Duhem inequality*. Thus *the dominant solutions represent the kinetic gas as a material in the sense of the modern thermomechanics of continua*. It is only in the thermo*dynamics*, which rests upon a specific principle of irreversibility, that there may be any difference.

Clearly the kinetic gas is homogeneous. I think that one could prove§ it to be a fluid in the sense of continuum mechanics, *viz*, the isotropy group

† In the Hilbert method, an initial-value problem for a system of partial differential equations is solved at each stage; in the Chapman-Enskog method, the fields ρ, θ, and $\dot{\mathbf{x}}$ are differentiated at each stage. Thus in both methods the basic fields are required in a neighborhood, not merely at a point.

‡ That the dominant solutions represent the kinetic gas as a material with memory in the sense of continuum mechanics seems first to have been remarked in Section 78 of my paper, "The mechanical foundations of elasticity and fluid dynamics," *J. Rational Mech. Anal.*, **1**:125–300 (1952); corrected reprint, *Continuum Mechanics*, vol. I, Gordon & Breach, New York, 1966. For more specific treatment, bringing out the precise nature of the memory effects represented, see the article by H. GRAD in the *Handbuch der Physik*, cited above.

§ For all Chapman-Enskog iterates so far exhibited explicitly, the proof is trivial.

of each of its constitutive equations is the full unimodular group. But *the kinetic gas is not a simple material.* To see this, it suffices to look at the formal iterates in the Chapman-Enskog process, which is set up expressly to find molecular-density functions F that *for all time* are functions of their own first five moments and the repeated space derivatives of these moments.

With respect to continuum theories as a whole, the position of the kinetic gas is summarized in the following table:

1° Exact agreement (all solutions): field equations for mass, linear momentum, rotational momentum, and energy.

2° Special features of the kinetic gas (all solutions):

$$p = \frac{k}{m}\rho\theta = \frac{2}{3}\rho\varepsilon, \qquad \gamma \equiv \frac{c_\pi}{c_v} = \frac{5}{3},$$

$$\mathbf{n}\cdot\mathbf{Pn} \geqq 0 \qquad \text{for all } \mathbf{n}.$$

3° Presently unsettled points: definition of η, relation of the H-theorem to the Clausius-Duhem inequality.

4° Constitutive equations (from dominant solutions, results partly conjectured): kinetic gas a non-simple homogeneous fluid, subject to special features listed under 2°.

That the kinetic gas is not a simple fluid has been used by some as an argument against modern continuum mechanics†, or at least as an argument against developing the theory of simple materials. They say, since so simple a theory of molecular structure as the kinetic theory of MAXWELL and BOLTZMANN implies that kinematic and thermal gradients of all orders may influence the pressure tensor and heating-flux vector, does it not follow that a theory leaving these higher gradients out of account has no applications at all? We could just as well say, since the kinetic theory allows us at most one independent linear viscosity, while

† The fact that the kinetic gas is not simple has been known since MAXWELL's day and is familiar to every student of continuum mechanics. See, for example, my paper written expressly because of that fact and no other: "A new definition of a fluid. II. The Maxwellian fluid," *J. Math. Pures Appl.*, (9) **30**:111–155 (1951), especially Section 18, and also the remark near the end of Section 26, *viz*, that the Navier-Stokes fluid is a common first approximation to infinitely many different theories of fluids, not necessarily simple.

A good deal of misunderstanding seems to arise from the differences in psychological motivation. Researchers in statistical or kinetic theories are inclined to claim a kind of universality for their own results and hence to presume in others like aspirations to empire. Modern continuum mechanics has been, from its start around 1945, frankly a theory of models. No-one, so far as I know, has *ever* claimed any universal truth for the theory of simple materials.

so simple a continuum theory as that of NAVIER and STOKES allows two and thus enables us to find that positive work is done in uniform expansion, as is almost invariably observed in practice, is it not futile to calculate higher and higher approximations in so degenerate a model as the kinetic gas? If the kinetic theory is so limited in first approximation, will not the "higher" approximations be still less applicable to reality?

The answer to both questions may be both yes and no, according to taste. For me, it is both times no. Both questions reflect a failure to come to grips with the real complications of nature. Beyond the easiest and long-mastered special cases, nature is too intricate for any inclusive theory. Different models have different uses; they emphasize different *aspects* of nature, often at the expense of leaving others altogether aside. Such is clearly the case in the kinetic and simple theories of fluids: One leaves out bulk viscosity, the other, some effects of non-homogeneity of the velocity and temperature-gradient fields; indeed, each leaves out much more, not the same for either. Each has its virtues—the kinetic theory, in particular, has a magnificent concreteness, allowing the theorist the joy of deriving specific numbers like 5:3 and 27:154 in return for grappling with sixfold integrals, a joy that is not spoilt by the merely physical fact that no experiment confirms these numbers.

One thing is certain: Merely negative criticism, on either side, gains nothing. Theories must be understood before they can be rejected. There is only one way to begin to understand a mathematical theory, even one with a physical purpose, and that is by proving mathematical theorems within it or about it. The kinetic theory has suffered at the hands of those kinetic theorists who indulge in orgies of "physical intuition", who revel in "interpretations" and "applications" without seeing that the theorems must be discovered and proved before they can be interpreted or applied. It was presented by MAXWELL as a mathematical theory; a mathematical theory it is; and as such it must be studied if we are ever to understand it.

In the two following lectures I will present some theorems of the kinetic theory. They are meager, but they are theorems.

Lecture 9

THE TREND TO EQUILIBRIUM ACCORDING TO THE KINETIC THEORY OF GASES

When we study the Maxwell-Boltzmann equation,

$$\partial_t F + \mathbf{v} \cdot \partial_{\mathbf{x}} F + \mathbf{b} \cdot \partial_{\mathbf{v}} F = \mathbb{C}[F], \tag{8.15}$$

where

$$\mathbb{C}[F] = \int\int w(F'F^{*\prime} - FF^*) \, dS \, d\mathbf{v}^*, \tag{8.17}$$

we distinguish between two kinds of solutions. The *general solutions* correspond to an initial-value problem. In principle, at least, they trace the evolution of the gas as some initial molecular-density function $F(\mathbf{x}, \mathbf{v}, 0)$ is converted, according to the Maxwell-Boltzmann equation, into the present molecular-density function $F(\mathbf{x}, \mathbf{v}, t)$. The various gross quantities defined from these solutions obey the field equations (8.22) of continuum mechanics; the kinetic equilibrium solutions lead to the same gross equations as those of Eulerian hydrostatics; but in general gas flows there is no reason to think that either the irreversibility law or any constitutive equation of continuum thermomechanics is satisfied. Solutions of a second and special kind, called *dominant solutions*, have been conjectured to exist, and formal procedures have been devised so as to find successive approximations to properties enjoyed by them. These solutions may or may not conform to the irreversibility law of continuum thermomechanics, but certainly they do obey specific constitutive relations defining a certain kind of non-simple material; indeed, the main aim of the classical approach to the kinetic theory is to calculate the response functions.

The relation between the general solutions and the dominant solutions—the former so far proved to exist only for a finite time interval and a special kind of molecules, the latter, so far proved to exist† only for one special flow, but both believed to exist, though in senses not yet specific—this relation offers the most tantalizing problem of the kinetic theory.

In the special case of equilibrium, this problem was first attacked by Maxwell. A gas in kinetic equilibrium obeys the equations of gross equilibrium in Eulerian hydrostatics, but the converse is not true. A gas

† I refer here to Section 34 of my paper, "On the pressures and the flux of energy in a gas according to Maxwell's kinetic theory, II," *J. Rational Mech. Anal.*, **5**:55–128 (1956), where the existence of "dominant parts" is proved for homothermal shearing. When, later in that paper (Section 42), I prove that "normal" solutions do not always exist, I refer to the more delicate fact described by (10.32), below.

grossly at rest does not have to be in kinetic equilibrium. Specifically, if $\mathbf{b} = \mathbf{0}$ and if the stream velocity, density, and temperature are uniform:

$$\dot{\mathbf{x}} = \text{const.}, \qquad \rho = \text{const.}, \qquad \theta = \text{const.} \tag{9.1}$$

in space and time, the molecular-density function $F(\mathbf{v}, t)$ need not be Maxwellian and generally is not. MAXWELL suggested, however, that as $t \to \infty$ the molecular density should approach that appropriate to kinetic equilibrium:

$$F(\mathbf{v}, t) \to F_{\mathrm{M}}(\mathbf{v}), \tag{9.2}$$

where F_{M} is the Maxwellian density that is determined uniquely by the constant parameters (9.1). That is, provided the molecular density be the same at all points of space at any one time, *any difference between gross equilibrium and kinetic equilibrium becomes insensible as time goes on.* Proof of this conjecture affords the famous problem of the *trend to equilibrium.*

While this problem is unsolved today, some steps have been made toward solving it, and the purpose of this lecture is to present in outline the full and rigorous solution in the special case when the molecules repel one another with a force inversely proportional to the fifth power of the distance. Such molecules, which have little to do with anything called a molecule in modern physics, being just as much mathematical constructs as are the Euler or Navier-Stokes equations, are called *Maxwellian.* In the earlier days great efforts were put out in the kinetic theory to get results for more realistic models of molecular interaction, but now that on the one hand the real molecules of the physicists grow more complicated every day, and on the other the difficulties of the kinetic theory itself even in this simplest, most oversimplified of cases are coming to be seen and faced, results obtained for Maxwellian molecules are not despised. Maxwellian molecules are useful today for the reason they were useful to MAXWELL, namely, that they allow us to calculate explicitly the effect of the collision operator on any polynomial function of the velocities and thus yield definite and rigorous if special results instead of the dubious approximations with which, alas, even today we must remain content if we insist upon using any other collision mechanism.

MAXWELL himself showed that in a Maxwellian gas grossly at rest, the pressure-deviator tensor and the heating-flux vector tend monotonically to zero with time. Specifically, he proved† that if F is spatially homogeneous, then

$$\begin{aligned} \mathbf{P}(t) - p\mathbf{1} &= (\mathbf{P}(0) - p\mathbf{1})e^{-t/\tau}, \\ \mathbf{q}(t) &= \mathbf{q}(0)e^{-\frac{2}{3}t/\tau}, \end{aligned} \tag{9.3}$$

† As noted by BOLTZMANN and POINCARÉ, MAXWELL's general equations for the third moments are not correct, but the error does not affect the specific results $(9.3)_2$ and (9.14).

where

$$\tau = \frac{\mu}{p},$$
$$\mu \equiv \frac{1}{3}\sqrt{\frac{2m}{G}}\frac{k\theta}{A_2}, \qquad A_2 = 1.3703 \ldots . \tag{9.4}$$

Here m is the mass of one molecule and G is the intermolecular force constant, force $= G/(\text{distance})^5$. In the next lecture I will show rigorously that μ deserves to be regarded as the shear viscosity of the gas, as indeed MAXWELL made plausible and everyone has long agreed, but in this lecture only the quantity τ, the *time of relaxation*, demands our attention as a gross parameter. Since

$$\tau = \frac{1}{3}\sqrt{\frac{2m}{G}} \cdot \frac{m}{\rho A_2}, \tag{9.5}$$

and since in this problem the density ρ is an assigned positive constant, τ also is an assigned positive constant here. MAXWELL's results (9.5) and (9.3) show that the rarer the gas is assumed to be, the longer is the time required for the pressure and the heating flux to fall to assigned fractions of their initial values. In any case, while in a gas grossly at rest indeed there may be at any one time arbitrary deviations from the results of Eulerian hydrostatics, such deviations decay monotonically to zero as time goes on, and they decay at specific rates determined by the two molecular constants and the assigned density.

From this sober work of MAXWELL have sprouted the daring if not fictional outbursts of relaxation times here, there, and everywhere. In contrast, MAXWELL's own analysis, while confined to a special problem, is above reproach, and his formulae (9.3) belong to that very select company, the exact theorems of the kinetic theory. The irreversibility of kinetic gas flows they prove beyond question, in that they afford a specific and easily grasped example of it; they suggest a trend to equilibrium, but they do not in themselves establish something so broad, since it remains possible that other expectations based on F do not approach the values appropriate to kinetic equilibrium. The kinetic gas is a complicated model; to rest content with those aspects that square with what we expect from elementary continuum mechanics, though a common sin among the "physically" inclined (who would be the last to suspect it in themselves), is to shut our eyes to its peculiar beauty.

When we come to demonstrate the trend to equilibrium, we need to face the equation of transfer. In place† of (8.21), which concerns func-

† With a now trivial exception, since the equation of linear momentum has been eliminated.

tions $f(\mathbf{v})$, we easily obtain an equation for the transfer of a function $f(\mathbf{c})$:

$$\mathbb{L}f = m\mathbb{C}f, \tag{9.6}$$

where, in indicial notation,

$$\begin{aligned} \mathbb{L}f &= (\rho\bar{f})^{\cdot} + \rho\bar{f}\dot{x}_{i,i} + \rho\overline{c_i f_{,j}}\dot{x}_{j,i} - \overline{f_{,j}}P_{ij,i} + (\rho\overline{c_i f})_{,i}, \\ \mathbb{C}f &= \int d\mathbf{v} \int d\mathbf{v}^* \int_0^\infty b\, db \int_0^{2\pi} w\, d\varepsilon \cdot (f^{*\prime} + f' - f^* - f)FF^*, \end{aligned} \tag{9.7}$$

and where b and ε are polar co-ordinates in the plane normal to $\mathbf{w}$ through $\mathbf{x}$.

To exploit this equation, MAXWELL suggested the use of spherical harmonics, and IKENBERRY has constructed a particular complete set which lends itself to easy calculation. First, let $Y_{i_1 i_2 \cdots i_n}(\mathbf{c})$ denote a homogeneous symmetric polynomial of degree n in the components of $\mathbf{c}$ such that the coefficient of $c_{i_1}c_{i_2} \cdots c_{i_n}$ is 1 and contraction on any pair of indices results in the zero polynomial:

$$\begin{aligned} Y(\mathbf{c}) &\equiv 1, \\ Y_i(\mathbf{c}) &\equiv c_i, \\ Y_{ij}(\mathbf{c}) &\equiv c_i c_j - \tfrac{1}{3}c^2\delta_{ij}, \\ Y_{ijk}(\mathbf{c}) &\equiv c_i c_j c_k - \tfrac{3}{5}c^2 c_{(i}\delta_{jk)}, \quad \ldots . \end{aligned} \tag{9.8}$$

Then set

$$Y_{2r|s} \equiv c^{2r} Y_s(\mathbf{c}), \tag{9.9}$$

where s stands for $i_1 i_2 \ldots i_s$. The spherical harmonics $Y_{2r|s}$ form a complete set of polynomials. The corresponding *spherical moments* $P_{2r|s}$ are defined by

$$P_{2r|s} \equiv \rho\overline{Y_{2r|s}} \equiv m\textstyle\int c^{2r} Y_s(\mathbf{c})F\, d\mathbf{c}. \tag{9.10}$$

The number $2r + s$ is the *order* of the spherical moment $P_{2r|s}$. The ordinary *moments* $p_{i_1 i_2 \cdots i_s}$ of F are defined as usual:

$$p_{i_1 i_2 \cdots i_s} \equiv \rho\overline{c_{i_1} c_{i_2} \cdots c_{i_s}} = m\textstyle\int c_{i_1} c_{i_2} \cdots c_{i_s} F\, d\mathbf{c}. \tag{9.11}$$

It is not hard to show that the spherical moments are linear combinations of the moments, and conversely. For example,

$$\begin{aligned} P_{0|ij} &= P_{ij} - p\delta_{ij}, \\ P_{2|0} &= 3p = 3kn\theta, \\ P_{0|ijk} &= p_{ijk} - \tfrac{3}{5}p_{ll(i}\delta_{jk)}, \\ P_{2|i} &= p_{ijj} = 2q_i, \\ P_{0|ijkl} &= p_{ijkl} - \tfrac{6}{7}p_{mm(ij}\delta_{kl)} + \tfrac{3}{35}p_{rrmm}\delta_{(ij}\delta_{kl)}, \quad \ldots . \end{aligned} \tag{9.12}$$

Moreover, if F is a Maxwellian molecular density, the spherical moments have the following special forms:

$$P_{2r|s} = P^{(0)}_{2r|s} \equiv \begin{cases} 0 & \text{if } s \neq 0, \\ (2r+1)!!\rho\left(\dfrac{p}{\rho}\right)^r & \text{if } s = 0. \end{cases} \tag{9.13}$$

By $(9.7)_1$ the effect of applying the differential operator $\mathbb{L}$ to $Y_{2r|s}$ leads to explicit expressions, if with some labor, and the forms of these are the same for all molecular models. For Maxwellian molecules MAXWELL himself observed that if g is a homogeneous polynomial of degree n, $\mathbb{C}g$ may be calculated explicitly as a bilinear function of the moments of orders j and $n - j$, summed over j, and that the form of this function is *the same for all solutions* F of the Maxwell-Boltzmann equation. Herein lies the mathematical simplicity of the Maxwellian molecular model. MAXWELL calculated explicit forms for the equations of transfer for the second and third moments, from which he obtained his celebrated results (9.3), and also one further relaxation theorem, which in the present notation we may write as

$$P_{0|ijk}(t) = P_{0|ijk}(0)e^{-\frac{3}{2}t/\tau}. \tag{9.14}$$

Thus MAXWELL found three times of relaxation: τ, $\frac{2}{3}\tau$, $\frac{3}{2}\tau$, for the moments of second and third orders. It is natural to ask whether the fourth moments relax, and if so, according to what law, and if exponentially, with what time or times of relaxation. When he tried to deal with this question, MAXWELL found that the equation of transfer for the fourth moments was too complicated for the answer to be seen, and the problem was left standing at this point for nearly eighty years.

IKENBERRY introduced his system of spherical moments just so as to obtain explicit equations of transfer in which a single moment of the order being studied appears, rather than the complicated linear combinations which occur with other choices of functions. He was able to prove the following ***fundamental theorem*†:**

$$m\mathbb{C}Y_{2r|s} = -C_{2r|s}P_{2r|s} + Q_{2r|s}, \tag{9.15}$$

where $C_{2r|s}$ is a positive constant, the value of which we shall shortly have to estimate, and where

$$Q_{2r|s} = \Sigma A_{r_1r_2s_1s_2}P_{2r_1|s_1}P_{2r_2|s_2},$$
$$2r_1 + s_1 \geqq 2r_2 + s_2 > 0, \qquad 2r_1 + s_1 + 2r_2 + s_2 = 2r + s. \tag{9.16}$$

† See Sections 7 and 13 of E. IKENBERRY & C. TRUESDELL, "On the pressures and the flux of energy in a gas according to Maxwell's kinetic theory, I," *J. Rational Mech. Anal.*, **5**:1–54 (1956).

Thus one spherical moment of the highest order, namely, $2r + s$, and one only, namely, $P_{2r|s}$ itself, appears on the right-hand side of (9.15). The proof of this theorem, while elementary, is formally too elaborate to present here.

The explicit collision integrals delivered by IKENBERRY's theorem enable us to prove the following ***theorem of the trend to equilibrium*** †: Let the gas be said to be in a state of *gross rest* if all moments of the molecular-density function F exist and are spatially homogeneous and once differentiable with respect to time. Then in a state of gross rest, *every moment relaxes exponentially to its equilibrium value, and no relaxation time is longer than that for the flux of energy,* viz $\frac{3}{2}\tau$.

First, in a state of gross rest the first five moments ρ, $\dot{\mathbf{x}}$, and θ are spatially homogeneous by assumption, and by the equations (8.22) expressing the balance of mass, momentum, and energy, they must be independent of time. Thus these constants determine a unique, time-independent Maxwellian density. Hence they determine unique equilibrium values of the spherical moments $P_{2r|s}$, namely, $P^{(0)}_{2r|s}$ as given by (9.13). The further moments are generally time-dependent. By $(9.7)_1$ and (9.10),

$$\mathbb{L}Y_{2r|s} = P'_{2r|s}, \tag{9.17}$$

where a prime denotes differentiation with respect to time. By (9.6) and (9.15), the functions $P_{2r|s}(t)$ satisfy the ordinary differential system

$$P'_{2r|s} = -\,C_{2r|s}P_{2r|s} + Q_{2r|s}, \tag{9.18}$$

linear in the moment of highest order occurring.

The first step in the proof of the theorem is to show that the general solution of (9.18) is of the form

$$P_{2r|s}(t) = \sum_{i=1}^{N} B_{2r|s|i}(t)\exp(-\lambda_i^{2r|s}\,t) + P^{(0)}_{2r|s}, \tag{9.19}$$

where the coefficient $B_{2r|s|i}(t)$ is a polynomial in t with coefficients which are functions of ρ, $\dot{\mathbf{x}}$, and θ; where N, the number of terms, depends on r and s; where the decay factor $\lambda_i^{2r|s}$ is positive; and where $P^{(0)}_{2r|s}$ is the equilibrium value of $P_{2r|s}$, given explicitly by (9.13). This formula shows that for large enough t the moment $P_{2r|s}(t)$ does approach its equilibrium value exponentially.

To prove (9.19), we first integrate (9.18) formally:

$$P_{2r|s}(t) = \exp(-C_{2r|s}t)[K_{2r|s} + \textstyle\int \exp(C_{2r|s}t)Q_{2r|s}(t)\,dt]. \tag{9.20}$$

† This theorem was obtained in Chapter VI of my paper, "On the pressures and the flux of energy in a gas according to Maxwell's kinetic theory, II," *op. cit.*

We then prove by induction that (9.20) has in fact the form (9.19). The first step is afforded by MAXWELL's results (9.3) and (9.14). Since $Q_{2r|s}$ is a bilinear combination of moments of orders lower than that of $P_{2r|s}$, the hypothesis of induction allows us to use (9.19) to evaluate $Q_{2r|s}$ explicitly by (9.16):

$$Q_{2r|s}(t) = \sum A_{r_1r_2s_1s_2}\Big[\sum_{i=1}^{N_1} B_{2r_1|s_1|i}(t) \exp\,(-\lambda_i^{2r_1|s_1}t) + P^{(0)}_{2r_1|s_1}\Big]\cdot$$
$$\Big[\sum_{j=1}^{N_1} B_{2r_2|s_2|j}(t) \exp\,(-\lambda_j^{2r_2|s_2}t) + P^{(0)}_{2r_2|s_2}\Big]. \qquad (9.21)$$

Hence

$$Q_{2r|s}(t) = \sum_{i=1}^{R} B_{2r|s|i}(t) \exp\,(\mu_{2r|s|i}t) + Q^{(0)}_{2r|s}, \qquad (9.22)$$

where $B_{2r|s|i}(t)$ is a polynomial, $\mu_{2r|s|i}$ is a positive constant, where r is a positive integer, and where $Q_{2r|s}(t) \to Q^{(0)}_{2r|s}$ as $t \to \infty$. Now from (9.21), the limit $Q^{(0)}_{2r|s}$ is the same bilinear combination of the $P^{(0)}_{2r|s}$ of lower order as $Q_{2r|s}(t)$ itself is of the $P_{2r|s}(t)$, and by the hypothesis of induction, these quantities $P^{(0)}_{2r|s}$ are not only the limits of $P^{(0)}_{2r|s}$ as $t \to \infty$ but also the equilibrium values of $P_{2r|s}$ for the given parameters ρ, $\dot{\mathbf{x}}$, and θ. Hence $Q^{(0)}_{2r|s}$ is, as the notation suggests, the equilibrium value of $Q_{2r|s}(t)$. This value, in turn, may be read off from the basic differential equation (9.18):

$$Q^{(0)}_{2r|s} = C_{2r|s}P^{(0)}_{2r|s}. \qquad (9.23)$$

If we substitute (9.22) and (9.23) in (9.20), we obtain

$$P_{2r|s}(t) = K_{2r|s} \exp\,(-C_{2r|s}t) + \sum_{i=1}^{R} C_{2r|s|i}(t) \exp\,(-\mu_{2r|s|i}t) + P^{(0)}_{2r|s}. \qquad (9.24)$$

where $K_{2r|s}$ is a constant of integration and $C_{2r|s|i}$ is a polynomial. Thus (9.19) is established.

To complete the proof of the theorem, we need to estimate the decay factors $\lambda_i^{2r|s}$, the reciprocals of which are the times of relaxation for the individual summands in (9.19). Specifically, we wish the least $\lambda_i^{2r|s}$ for given r and s, since this one corresponds to the longest time of relaxation and hence yields the term that ultimately dominates the sum as $t \to \infty$. Setting

$$\lambda^{2r|s} \equiv \operatorname*{Min}_i \lambda_i^{2r|s}, \qquad (9.25)$$

we shall prove that

$$\tau\lambda^{2r|s} \geqq \tfrac{2}{3} \qquad \text{if } 2r + s \geqq 3. \qquad (9.26)$$

MAXWELL's results $(9.3)_2$ and (9.14) may be put in the forms

$$\tau\lambda^{2|1} = \tfrac{2}{3}, \qquad \tau\lambda^{3|0} = \tfrac{3}{2}, \tag{9.27}$$

and an easy explicit calculation yields

$$\tau\lambda^{4|0} = \tfrac{2}{3}, \qquad \tau\lambda^{2|2} = 1, \qquad \tau\lambda^{0|4} = 2. \tag{9.28}$$

Thus the assertion is true when $2r + s = 3$ and 4. We shall prove by induction that it is true in general. The exponents $\mu_{2r|s|i}$ in (9.24) are those first occurring in (9.22). The proof of (9.22) shows that these numbers are the sums of pairs of $\lambda_i^{2r|s}$ of lower orders $2r + s$. By the hypothesis of induction, these sums cannot be less than $\frac{2}{3}/\tau$. Hence (9.24) yields at once

$$\lambda^{2r|s} \geqq \operatorname{Min}\left[C_{2r|s}, \frac{2}{3\tau}\right]. \tag{9.29}$$

The proof is thus reduced to showing that

$$C_{2r|s} \geqq \frac{2}{3\tau}. \tag{9.30}$$

This last inequality is true, and the lower bound is the best possible since

$$C_{2|1} = \frac{2}{3\tau}, \tag{9.31}$$

this being again MAXWELL's result $(9.3)_2$ in another notation. To prove the inequality (9.30), we need an explicit formula for the coefficient $C_{2r|s}$ in IKENBERRY's theorem (9.15). In fact

$$C_{2r|s} = 4\pi n\sqrt{\frac{G}{2m}}\int_0^\pi [1 + \delta_{r0}\delta_{s0} - \cos^{2r+s}\tfrac{1}{2}\phi P_s(\cos\tfrac{1}{2}\phi) - \sin^{2r+s}\tfrac{1}{2}\phi P_s(\sin\tfrac{1}{2}\phi)]\alpha\, d\alpha, \tag{9.32}$$

where $P_s(\cdot)$ is the Legendre polynomial of order s and where α and ϕ are related by the parametric equations

$$\alpha^4 = 2\cot^2 2\theta, \qquad \phi = 2\sqrt{\cos 2\theta}\, K(\sin\theta), \tag{9.33}$$

$K(\cdot)$ being the complete elliptic integral of the second kind. Hence, clearly,

$$C_{2r|s} \geqq C_{q|0}, \qquad q = 2r + s, \qquad C_{q|0} \equiv 4\pi n\sqrt{\frac{G}{2m}}\int_0^\pi (1 + \delta_{q0} - \cos^q\tfrac{1}{2}\phi - \sin^q\tfrac{1}{2}\phi)\alpha\, d\alpha. \tag{9.34}$$

Since the integrand is an increasing function of q at all points in the interval $0 < \phi < \pi$,

$$C_{q|0} > C_{q'|0} \qquad \text{if } q > q'. \tag{9.35}$$

Hence if $2r + s \geqq 4$,

$$C_{2r|s} \geqq C_{4|0} = \frac{2}{3\tau}, \tag{9.36}$$

so the desired inequality (9.30) is established.

The proof just given rests heavily on IKENBERRY's theorem, which in turn is peculiar to Maxwellian molecules. We have presently no hint toward proving the corresponding theorem for other laws of interaction; indeed, great effort has been made with only slight success†. With modern analytical tools, a general theorem of the trend to equilibrium should be proved soon, but we can scarcely expect the approach to be so simple as to allow us to delimit it by times of relaxation, as I have done here for the gas whose molecules are Maxwellian.

† T. CARLEMAN, *Problèmes mathématiques dans la théorie cinétique des gaz*, Almqvist & Wiksells Bocktryckeri, Uppsala, 1957. As noted by the editors of this posthumous work, there are major gaps in the argument, so that the only thing really established is the result which is contained also in an earlier paper of CARLEMAN, "Sur la théorie de l'équation intégrodifférentiel de Boltzmann," *Acta Math.*, **60**:91–146 (1933), where densities isotropic in **v** as well as spatially homogeneous were considered for a gas whose molecules are elastic spheres.

For a certain mollified Maxwell-Boltzmann equation with bounded collision cross section, a certain weak trend of spatially homogeneous solutions to equilibrium has been proved by D. MORGENSTERN, "Analytical studies related to the Maxwell-Boltzmann equation," *J. Rational Mech. Anal.*, **4**:533–555 (1955).

A similar result for KAC's one-dimensional caricature of a gas has been established by H. P. MCKEAN, JR., "Speed of approach to equilibrium for Kac's caricature of a Maxwellian gas," *Arch. Rational Mech. Anal.*, **21**:341–367 (1966).

From these works it seems that the analytical problem for any case other than that described in this lecture must be extremely difficult.

LECTURE 10

KINETIC AND CALORIC DISSIPATION IN SIMPLE SHEARING OF A MAXWELLIAN GAS

In the preceding lecture "gross rest" was defined as a condition of the kinetic gas in which the moments of all orders are spatially homogeneous. Considering the case where there is no body force, we saw that the equations of transfer for the moments of the molecular-density function reduced for a Maxwellian gas to a system of ordinary differential equations which could be integrated explicitly, one after the other. Gross rest, however, is not the only condition in which, for a Maxwellian gas, the moments of order greater than n in the equation of transfer for the nth moments drop out, so that the system of differential equations satisfied by the moments of lower order becomes a determinate one. Since the equations of transfer are of the form

$$\mathbb{L}f = m\mathbb{C}f, \tag{9.6}$$

where

$$\mathbb{L}f = (\rho\bar{f})^{\cdot} + \rho\bar{f}\dot{x}_{k,k} + \rho\overline{c_k f_{,l}}\dot{x}_{l,m} - \overline{f_{,k}}P_{lk,l} + (\rho\overline{c_k f})_{,k}, \tag{9.7$_1$}$$

if we assume that the moments $p_{i_1,i_2 \ldots i_r}$ are spatially homogeneous when $r > r_0$, say, then the last term vanishes when $r \geqq r_0$, so that

$$\mathbb{L}c_{i_1}c_{i_2} \cdots c_{i_r} = \text{a linear combination of moments of orders } 1, 2, \ldots, r, \tag{10.1}$$

provided $r \geqq r_0$. That is, the equations of transfer for the moments of orders $0, 1, 2, \ldots, r_0$ form a system of partial differential equations by themselves. If these can be solved, the solutions may be put into the equations of the moments of orders $r_0 + 1$, $r_0 + 2$, ... Therefore, the forward coupling of $\mathbb{L}c_{i_1}c_{i_2} \cdots c_{i_r}$ is annulled for these solutions; for Maxwellian molecules, there is no forward coupling in $\mathbb{C}c_{i_1}c_{i_2} \cdots c_{i_r}$ for any solution. Thus for these particular solutions, the formidable forward coupling of the equations of transfer is outflanked because it breaks off after a finite number of occurrences.

Whether there are any solutions of the type just assumed remains to be verified. The preceding lecture concerned the case when $r_0 = 0$ and exhibited all solutions corresponding to it. In this lecture I shall show that there are indeed solutions when $r_0 = 1$.

The equations of transfer for the moments of orders 0, 1, and 2 are the ordinary equations of balance in continuum mechanics (8.22), as shown in Lecture 8. Here I write them in the slightly different forms

$$\begin{aligned} \dot\rho &= -\rho \operatorname{div} \dot{\mathbf{x}}, \\ \ddot{\mathbf{x}} &= -\frac{1}{\rho} \operatorname{div} \mathbf{P} + \mathbf{b}, \\ \tfrac{3}{2}\dot p &= -\tfrac{5}{2}p \operatorname{div} \dot{\mathbf{x}} - \operatorname{tr}(\mathbf{PD}) - \operatorname{div} \mathbf{q}. \end{aligned} \tag{10.2}$$

When $\mathbf{b} = \mathbf{0}$ and when p, $\mathbf{P}$, and $\mathbf{q}$ are spatially homogeneous, these equations reduce to

$$\begin{aligned} \dot\rho &= -\rho \operatorname{div} \dot{\mathbf{x}}, \\ \ddot{\mathbf{x}} &= \mathbf{0}, \\ \tfrac{3}{2}p' &= -\tfrac{5}{2}p \operatorname{div} \dot{\mathbf{x}} - \operatorname{tr}(\mathbf{PD}), \end{aligned} \tag{10.3}$$

where the prime denotes differentiation of a function of time alone. Thus far, no restriction has been laid upon the velocity field $\dot{\mathbf{x}}$ beyond that expressed by $(10.3)_2$, namely, it is accelerationless, so that each fluid particle† travels along some straight line at constant speed. While the general solution of the system (10.3) seems not to be known, an important class of particular solutions is furnished by *steady homogeneous velocity fields:*

$$\dot{\mathbf{x}} = \mathbf{Gp}, \qquad \begin{aligned} \mathbf{G} &= \text{const.}, \\ \mathbf{p} &= \text{position vector}. \end{aligned} \tag{10.4}$$

In order that (10.4) satisfy $(10.3)_2$, it is necessary and sufficient that

$$\mathbf{G}^2 = \mathbf{0}. \tag{10.5}$$

In the special case when $\operatorname{tr} \mathbf{G} = 0$, $(10.3)_1$ becomes

$$\dot\rho = 0: \tag{10.6}$$

The flow is isochoric. We shall consider here only the still more special case when in fact‡

$$\rho = \text{const. in space and time.} \tag{10.7}$$

It follows that

$$\theta = \text{const. in space:} \tag{10.8}$$

The flow is homothermal. The only one of equations (10.3) not trivially satisfied is the energy equation $(10.3)_3$, which yields a connection between

† Of course, the term "particle" here is used in the sense of continuum mechanics, referring to the mean velocity $\dot{\mathbf{x}}$ of the gas molecules, as defined by $(8.3)_2$.

‡ This lecture presents some of the results from Chapter V of my paper, "On the pressures and the flux of energy in a gas according to Maxwell's kinetic theory, II," *J. Rational Mech. Anal.*, **5**:55–128 (1956). The existence of this class of solutions was discovered at the same time by Galkin, «Об одном решении кинетического уравнения Больцмана,» *Prikl. Mat. Mekh.*, **20**:445–446 (1956).

$p(t)$ and $\mathbf{P}(t)$:

$$\tfrac{3}{2}p' = -\operatorname{tr}(\mathbf{PD}), \qquad (10.9)$$

where $2\mathbf{D} = \mathbf{G} + \mathbf{G}^T$.

So as to picture the solution easily, we shall consider, finally, only the very special case of simple shearing (Figure 10), although results of the same type hold with no restriction beyond (10.4), (10.5), and (10.6). In our special case

$$\dot{x} = Ky, \quad \dot{y} = 0, \quad \dot{z} = 0, \quad K = \text{const.},$$
$$[\mathbf{G}] = \begin{Vmatrix} 0 & K & 0 \\ 0 & 0 & 0 \\ 0 & 0 & 0 \end{Vmatrix}, \quad [2\mathbf{D}] = \begin{Vmatrix} 0 & K & 0 \\ K & 0 & 0 \\ 0 & 0 & 0 \end{Vmatrix}. \qquad (10.10)$$

The flow may be regarded as induced† by the sliding of a plate at $y = 0$. The temperature is uniform at each time, and there is no heating flux.

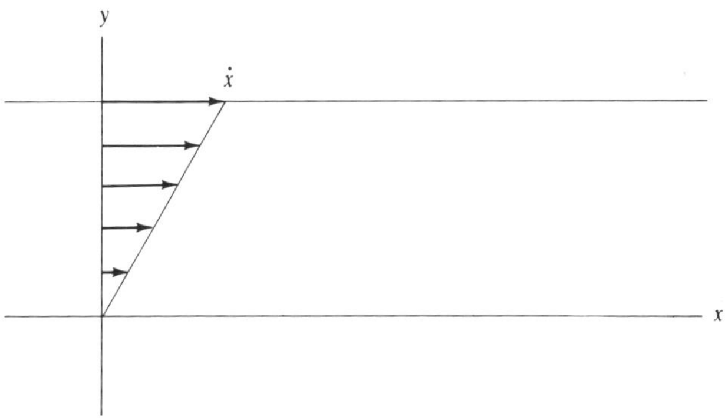

FIGURE 10. Velocity profile in simple shearing.

Since work is being done, the gas becomes steadily hotter, and consequently, since the density is constant in time, the pressure grows infinitely large as $t \to \infty$. The rate at which it does so is restricted by the energy equation (10.9), which assumes the form

$$p' + \tfrac{2}{3}KP_{xy} = 0. \qquad (10.11)$$

By hypothesis, P_{xy} is a function of time only, but what function it is depends upon the constitutive equations of the gas‡.

† In this interpretation, the fluid adheres to the plates. Of course the same flow is consistent also with a boundary condition prescribing a slip velocity in terms of θ, K, *etc.*

‡ The problem treated here seems not to be very well known even in the Navier-Stokes theory, for which

$$P_{xy} = \mu(p, \theta)K,$$

so that (10.11) becomes

$$p' + \tfrac{2}{3}K^2 f(p) = 0,$$

Thus far, we have made no use of the specific nature of the kinetic theory. While the constitutive equations implied by the kinetic theory are not known in general, they can be found for the particular flow we are considering, found exactly and for all possible initial conditions. It is instructive first to look at the general equations of transfer for the second and third moments for a gas of Maxwellian molecules†:

$$\begin{aligned} -\frac{p}{\mu} P_{0|km} &= \dot{P}_{km} + P_{km}\dot{x}_{q,q} + 2P_{q(k}\dot{x}_{m),q} + p_{kmq,q}, \\ -\frac{p}{2\mu}(3p_{kms} - 2q_{(k}\delta_{ms)}) &= \dot{p}_{kms} + p_{kms}\dot{x}_{r,r} + p_{kmsr,r} \\ &\quad - \tfrac{3}{2}P_{(km}P_{s)r,r} + 3p_{r(km}\dot{x}_{s),r}, \end{aligned} \tag{10.12}$$

where the superimposed dot denotes the material derivative with respect to the velocity field $\dot{\mathbf{x}}$; where $\mathbf{P}_0$ is the pressure-deviator tensor; and where μ is the formal abbreviation we encountered in the previous lecture:

$$\mu \equiv \frac{1}{3}\sqrt{\frac{2m}{G}\frac{k\theta}{A_2}}, \qquad A_2 = 1.3703 \ldots . \qquad (9.4)_{2,3}$$

For the special problem of homothermal shearing, not only do the forward-coupling terms vanish, but also $(10.12)_1$ when set alongside (10.11) yields a system of ordinary differential equations for determination of the pressure p and the pressure-deviator tensor $\mathbf{P}_0$ from their initial values and from the parameters defining the problem. These parameters are, first, the *time of relaxation* τ:

$$\tau \equiv \frac{\mu}{p} = \frac{1}{3}\sqrt{\frac{2m}{G}}\cdot\frac{m}{\rho A_2}, \qquad (9.4)_1,\ (9.5)$$

---

the integral of which is

$$\int \frac{dp}{f(p)} = -\tfrac{2}{3}K^2 t.$$

For example, for a perfect Navier-Stokes gas with viscosity proportional to the temperature, we obtain

$$\frac{p}{p_0} = \frac{\theta}{\theta_0} = \exp\left(\tfrac{2}{3}\mathfrak{T}^2\frac{t}{\tau}\right),$$

where $\tau \equiv \mu/p = (\text{const.})/\rho$ and $\mathfrak{T} \equiv \tau K$. Thus p, θ, μ, and $|P_{xy}|$ all grow exponentially with time as a result of the work done on the gas. Only in the rate and nature of that growth do the results we obtain below for the kinetic gas differ from those for the Navier-Stokes gas.

† While MAXWELL did not publish all the terms in these equations, his explanations make it clear that he had them, though with some slips in calculation. Of course they are more easily gotten by use of IKENBERRY's spherical moments (9.10).

and, second, the dimensionless *truncation number* $\mathfrak{T}$:

$$\mathfrak{T} \equiv \frac{\mu K}{p} = \tau K. \tag{10.13}$$

Small values of $\mathfrak{T}$ correspond to a shearing that is small in comparison with the reciprocal time of relaxation of the gas at the density assigned for the problem. If, on the other hand, we prefer to give K a single fixed value, then by (9.5) the limit as $\mathfrak{T} \to \infty$ corresponds to ultimate rarefaction of the gas.

The final system of ordinary differential equations to be solved is

$$\begin{aligned} \tau p' + \tfrac{2}{3}\mathfrak{T} P_{xy} &= 0, \\ \tau P'_{xy} + P_{xy} + \mathfrak{T} p + \mathfrak{T} P_{0|yy} &= 0, \\ \tau P'_{0|yy} + P_{0|yy} - \tfrac{2}{3}\mathfrak{T} P_{xy} &= 0, \\ \tau P'_{0|xx} + P_{0|xx} + \tfrac{4}{3}\mathfrak{T} P_{xy} &= 0, \\ \tau P'_{xz} + P_{xz} &= 0, \\ \tau P'_{yz} + P_{yz} &= 0. \end{aligned} \tag{10.14}$$

When $\mathfrak{T} = 0$, these reduce to the equations for the second moments in the problem of the trend to equilibrium, solved in the preceding lecture. While we were there interested in the behavior of all the moments as $t \to \infty$, here we fasten our attention upon the pressure tensor as a function of the time and the truncation number. While in the problem of the trend to equilibrium we proved what we expected, namely, that gross rest is ultimately indistinguishable from kinetic equilibrium, here we certainly do not expect a trend to equilibrium, since work must be done continually upon the gas in order to effect the motion. While we do not know just what to expect, we may phrase a similar question: Does the pressure tensor **P** approach with time the value it would assume right away according to some simple theory of continuum mechanics such as the Navier-Stokes theory of linearly viscous fluids? More generally, what relations hold between the results of the kinetic theory and those of modern continuum mechanics in this particular flow?

The last two of Equations (10.14) show that the shear stresses on the planes of flow relax just as in a state of gross rest:

$$\begin{aligned} P_{xz}(t) &= P_{xz}(0)e^{-t/\tau}, \\ P_{yz}(t) &= P_{yz}(0)e^{-t/\tau}. \end{aligned} \tag{10.15}$$

The remaining four components of the pressure tensor satisfy the linear system $(10.14)_{1,2,3,4}$. There are solutions proportional to $\exp{(\chi t/\tau)}$ provided the growth modulus χ satisfy the cubic equation

$$\chi(\chi + 1)^2 = \tfrac{2}{3}\mathfrak{T}^2. \tag{10.16}$$

This equation has one positive root $\mathfrak{r}$ and a pair of conjugate complex roots $\mathfrak{R} \pm i\mathfrak{J}\mathfrak{T}$, where

$$\mathfrak{R} = -(1 + \tfrac{1}{2}\mathfrak{r}), \qquad \mathfrak{J}^2\mathfrak{T}^2 = \mathfrak{r}(1 + \tfrac{3}{4}\mathfrak{r}). \tag{10.17}$$

The positive root $\mathfrak{r}$ is a monotonically increasing function of $\frac{2}{3}\mathfrak{T}^2$ (see

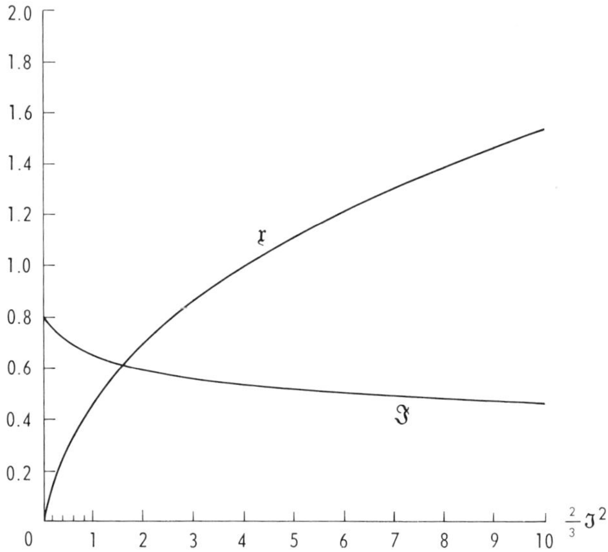

FIGURE 11. Growth modulus $\mathfrak{r}$ and frequency modulus $\mathfrak{J}$ for shearing of a Maxwellian gas.

Figure 11). The general solution of the initial-value problem is

$$\frac{p}{p_0} = Ae^{\mathfrak{r}t/\tau} + e^{-(1+\frac{1}{2}\mathfrak{r})t/\tau}[B\cos\mathfrak{J}Kt + C\sin\mathfrak{J}Kt],$$

$$\begin{aligned}\frac{P_{xy}}{p_0} = \frac{1}{\frac{2}{3}\mathfrak{T}}\{-A\mathfrak{r}e^{\mathfrak{r}t/\tau} + e^{-(1+\frac{1}{2}\mathfrak{r})t/\tau} \\ \cdot[(\{1 + \tfrac{1}{2}\mathfrak{r}\}B - \mathfrak{J}\mathfrak{T}C)\cos\mathfrak{J}Kt \\ + (\{1 + \tfrac{1}{2}\mathfrak{r}\}C + \mathfrak{J}\mathfrak{T}B)\sin\mathfrak{J}Kt]\},\end{aligned} \tag{10.18}$$

$$\begin{aligned}\frac{P_{0|yy}}{p_0} = \frac{-1}{\mathfrak{r}+1}\Bigg\{A\mathfrak{r}e^{\mathfrak{r}t/\tau} + e^{-(1+\frac{1}{2}\mathfrak{r})t/\tau}\Bigg[\left(\{\tfrac{3}{2} + \mathfrak{r}\}B + \frac{\mathfrak{J}\mathfrak{T}}{\mathfrak{r}}C\right)\cos\mathfrak{J}Kt \\ + \left(\{\tfrac{3}{2} + \mathfrak{r}\}C - \frac{\mathfrak{J}\mathfrak{T}}{\mathfrak{r}}B\right)\sin\mathfrak{J}Kt\Bigg]\Bigg\},\end{aligned}$$

where the constants A, B, and C are determined from the initial pressures p_0, $P_{xy}(0)$, and $P_{0|yy}(0)$ as follows:

$$\begin{aligned} 1 &= A + B, \\ \frac{\frac{2}{3}\mathfrak{T} P_{xy}(0)}{p_0} &= -A\mathfrak{r} + (1 + \tfrac{1}{2}\mathfrak{r})B - \mathfrak{Z}\mathfrak{T}C, \\ \frac{-(\mathfrak{r}+1)P_{0|yy}(0)}{p_0} &= A\mathfrak{r} + (\tfrac{3}{2} + \mathfrak{r})B + \frac{\mathfrak{Z}\mathfrak{T}}{\mathfrak{r}}C. \end{aligned} \tag{10.19}$$

The remainder of the system of pressures may be obtained from two simple relaxation theorems which follow easily from (10.14):

$$\begin{aligned} P_{yy}(t) - P_{zz}(t) &= [P_{yy}(0) - P_{zz}(0)]e^{-t/\tau}, \\ P_{0|xx}(t) + 2P_{0|yy}(t) &= [P_{0|xx}(0) + 2P_{0|yy}(0)]e^{-t/\tau}. \end{aligned} \tag{10.20}$$

From these results we see at once that a part of the solution oscillates in time but is ultimately damped out. There is thus *a trend in time*, which we shall study further. To ease the description, the part of the solution that is monotonically increasing, since it becomes ultimately far larger than the *oscillating part* p^{O}, $\mathbf{P}^{\mathrm{O}}$, may be called the *dominant part* p^{D}, $\mathbf{P}^{\mathrm{D}}$:

$$p = p^{\mathrm{D}} + p^{\mathrm{O}}, \qquad \mathbf{P} = \mathbf{P}^{\mathrm{D}} + \mathbf{P}^{\mathrm{O}}. \tag{10.21}$$

This example shows that the behavior described by MAXWELL's relaxation theorem for gross rest need not take place in more general conditions. In gross rest, part of the system of pressures dominates because it stays constant, while the remainder decays steadily to zero. Here, on the contrary, the dominant part grows in time, while the damped part oscillates. The two cases are of one kind only in the clear separation of the pressures into damped and undamped parts.

Even the special case furnished by MAXWELL's relaxation theorem shows that it is only the dominant parts we can expect to compare with the results of the theory of simple materials. From (10.18) we may read off the values of these dominant parts:

$$\begin{aligned} \frac{p^{\mathrm{D}}}{p_0} &= Ae^{\mathfrak{r}t/\tau}, \\ \frac{P_{xy}{}^{\mathrm{D}}}{p_0} &= \frac{-A\mathfrak{r}}{\frac{2}{3}\mathfrak{T}}e^{\mathfrak{r}t/\tau}, \\ \frac{P_{yy}{}^{\mathrm{D}} - p^{\mathrm{D}}}{p_0} &= \frac{-A\mathfrak{r}}{\mathfrak{r}+1}e^{\mathfrak{r}t/\tau}, \\ P_{zz}{}^{\mathrm{D}} &= P_{yy}{}^{\mathrm{D}}, \\ P_{xx}{}^{\mathrm{D}} &= 3p^{\mathrm{D}} - 2P_{yy}{}^{\mathrm{D}}. \end{aligned} \tag{10.22}$$

All the dominant pressures are time-dependent, and all depend upon initial conditions, but the rates of growth of all are the same, and likewise their dependence upon the initial conditions is only through the common factor of proportionality A.

In modern continuum mechanics a rather general kind of fluid, the *simple fluid* of NOLL, has been introduced and studied. In particular, the stress system in such a fluid subjected to a steady *viscometric flow* has been determined explicitly†. Simple shearing is included as a special case in the results‡, which show that

$$\begin{aligned} P_{xy} &= -\sigma_0(K;\rho,\theta), \\ P_{yy} - P_{zz} &= -\sigma_1(K;\rho,\theta), \\ P_{xx} - P_{zz} &= -\sigma_2(K;\rho,\theta), \end{aligned} \tag{10.23}$$

where σ_0, σ_1, and σ_2 are called the *viscometric functions*. As a function of K, σ_0 is odd, while σ_1 and σ_2 are even; moreover, σ_0 is non-negative. Beyond that, the theory of simple fluids allows σ_0, σ_1, and σ_2 to be arbitrary. These are the results with which we are to compare (10.22), the solution for homothermal shearing according to the kinetic theory for Maxwellian molecules.

If we use $(10.22)_1$ to evaluate p^{D}, we can divide the rest of Equations (10.22) by it and so obtain time-independent ratios $P_{km}{}^{\mathrm{D}}/p^{\mathrm{D}}$. These are of just the kind delivered by continuum mechanics. Indeed, since

$$p = \frac{k}{m}\rho\theta, \tag*{(8.8)$_2$}$$

we can define a dominant temperature θ^{D} by the formula

$$p^{\mathrm{D}} = \frac{k}{m}\rho\theta^{\mathrm{D}}, \tag{10.24}$$

where p^{D} is given by $(10.22)_1$. Thus dependence upon p^{D} amounts to a special dependence upon ρ and θ^{D}. If we set

$$\begin{aligned} \sigma_0(K;\rho,\theta) &= \tfrac{3}{2}p^{\mathrm{D}}\frac{\mathfrak{r}}{\mathfrak{T}}, \\ \sigma_1(K;\rho,\theta) &= 0, \\ \sigma_2(K;\rho,\theta) &= 3p^{\mathrm{D}}\frac{\mathfrak{r}}{1+\mathfrak{r}}, \end{aligned} \tag{10.25}$$

† For an exposition, see Chapter E of *The Non-Linear Field Theories of Mechanics*, in FLÜGGE's *Handbuch der Physik*, vol. III/3, Springer-Verlag, 1965.

‡ That is, for the form of the stress system. Dynamical conditions have been worked out only for incompressible fluids, so the results (10.22) cannot be read off from the known theorems on the viscometry of simple fluids.

and then recall that $\mathfrak{r}$ is a function of the truncation number $\mathfrak{T}$ alone and that $\mathfrak{T}$ is proportional to the shearing K, namely,

$$\mathfrak{T} \equiv \frac{\mu K}{p} = \tau K = \frac{1}{3}\sqrt{\frac{2m}{G}} \cdot \frac{m}{\rho A_2} K, \qquad (9.5), (10.13)$$

we see that the relations (10.25) are included as special cases in (10.23). That is, *in homothermal shearing a Maxwellian gas obeys, ultimately, the constitutive equation of a particular simple fluid.* (Of course, in this flow the thermokinetic gradients of order greater than 1 all vanish, so no distinction between a simple material and a material of higher grade could be observed here.)

The nature of this particular simple fluid deserves to be remarked. First, it is easy to show from (10.16) that

$$\mathfrak{r} = \tfrac{2}{3}\mathfrak{T}^2(1 - \tfrac{4}{3}\mathfrak{T}^2) + O(\mathfrak{T}^6) \qquad \text{as } \mathfrak{T} \to 0. \qquad (10.26)$$

Hence $(10.25)_1$ yields

$$\begin{aligned}\sigma_0(K; \rho, \theta) &= p^{\mathrm{D}}\mathfrak{T} + O(\mathfrak{T}^2),\\ &= \frac{p^{\mathrm{D}}}{p}\mu K + O(\mathfrak{T}^2) \qquad \text{as } \mathfrak{T} \to 0. \qquad (10.27)\end{aligned}$$

Since, ultimately, $p^{\mathrm{D}}/p \to 1$ as $t \to \infty$ for any fixed $\mathfrak{T}$, this last result shows that for sufficiently slow shearings in a sufficiently dense gas the shear-viscosity function is approximated by μK, the result for a linearly viscous fluid with shear viscosity μ. Thus μ, which in these lectures was introduced by the formal definition $(9.4)_2$, turns out to be the shear viscosity of the fluid in one of the classical senses of the term†.

Second, the fact that $\sigma_1(K; \rho, \theta) = 0$ means that all planes parallel to the direction of flow suffer, ultimately, equal normal pressures. A continuum theory in which this is true is said to obey "WEISSENBERG's assertion".

Third, the nature of the viscometric functions themselves is of great

† The reader not familiar with modern continuum mechanics may need to be reminded that there are several different concepts of shear viscosity. The one used here refers to viscometric flows alone. We have shown that *for homothermal shearing* μ as defined by $(9.4)_2$ is the linear shear viscosity in the sense of the modern theory of viscometry. We have *not* shown that any viscometric flows *other* than homothermal shearing are possible for the Maxwellian gas; thus, *a fortiori*, we have *not* shown that μ, or more generally the functions σ_0, σ_1, and σ_2 occurring in (10.23), have any particular interpretation in such more general flows. Still less do the results here have bearing on the second common approach to the concept of viscosity, that which regards the Navier-Stokes equations as governing *all* flows for which $|\mathbf{D}|$ is small enough. This second concept, of course, is employed in the discussion of bulk viscosity given above in Lecture 8.

interest. If we divide them by p^{D}, we obtain reduced viscometric functions†, that is, dimensionless functions of the dimensionless truncation number $\mathfrak{T}$ alone:

$$
\begin{aligned}
V(\mathfrak{T}) &\equiv \frac{\sigma_0(K;\, \rho,\, \theta)}{p^{\mathrm{D}}} = \frac{3}{2}\frac{\mathfrak{r}}{\mathfrak{T}},\\
N_1(\mathfrak{T}) &\equiv \frac{\sigma_1(K;\, \rho,\, \theta)}{p^{\mathrm{D}}} = 0, \qquad (10.28)\\
N_2(\mathfrak{T}) &\equiv \frac{\sigma_2(K;\, \rho,\, \theta)}{p^{\mathrm{D}}} = \frac{3\mathfrak{r}}{1+\mathfrak{r}}.
\end{aligned}
$$

That the shear viscosity μ of a Maxwellian gas is independent of ρ and proportional to θ, was one of the most striking early products of the kinetic theory, discovered by MAXWELL himself. In the present notation, this result is expressed in the form $V'(0) = 1$. We see that in fact a far more

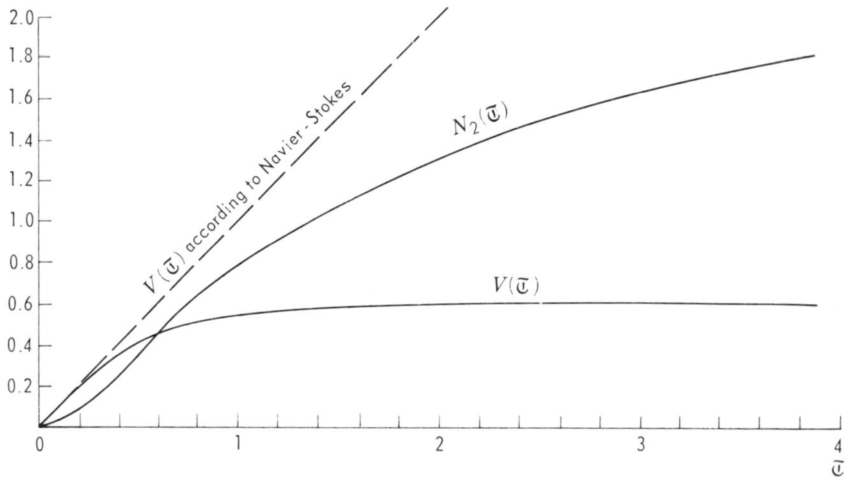

FIGURE 12. Reduced viscometric functions for the Maxwellian gas (the asymptotic value 3 for $N_2(\mathfrak{T})$ is approached too slowly to be discernible on this drawing).

general statement of the latter kind holds, valid for all densities and shearings: *All three reduced viscometric functions are independent of the temperature.* The former property discovered by MAXWELL, however, is valid only in the limit when $\mathfrak{T}$ is small, that is, for high densities or small shearings. *The reduced viscometric functions depend on K and ρ only through the dimensionless ratio $\mathfrak{T}$.* That is, *they are functions only of the ratio of the shearing to the reciprocal of the time of relaxation.* The reduced shear-

† These reduced functions are constant multiples of those discussed by NOLL & me in Section 108 of our treatise.

viscosity function V, while linear for small $\mathfrak{T}$, falls away to a maximum value $\sqrt{\frac{3}{8}}$ at $\mathfrak{T} = \sqrt{6}$, decreasing thereafter to 0 as $\mathfrak{T} \to \infty$. The reduced normal stress function N_2 vanishes, of course, when $\mathfrak{T} = 0$; it increases monotonically to the asymptotic value 3 as $\mathfrak{T} \to \infty$. Thus the larger is $\mathfrak{T}$, the less important is the effect of shear viscosity relative to the difference of normal pressures. These results are illustrated in Figure 12. At the risk of seeming to wander into fields ruled by dangerous looseness of thought, I may say that the Maxwellian gas at low densities or high shearings is "non-Newtonian".

We may summarize the results so far as showing that while the Navier-Stokes theory gives only a crude first approximation to the phenomena occurring in homothermal shearing of a Maxwellian gas, these are included as a very special case in the modern theory of viscometry.

The kinetic theory is mathematically so difficult that to treat almost any other problem, methods of approximation seem to be needed, and, as is often the case when approximation is really important, nothing has been proved, though much has been conjectured or even asserted. The methods of approximation can be treated pragmatically by trying them out on the example of homothermal shear flow, since it is the only dissipative flow so far known to furnish an exact solution. The method of approximation generally favored by physicists is that of CHAPMAN and ENSKOG. While it is very complicated, the present example can be treated by a simpler method called *atemporal Maxwellian iteration*†, which seems to yield for Maxwellian molecules the same results as the Chapman-Enskog process. This method seeks iterates $\mathbf{P}^{\{n\}}$ to the pressure tensor $\mathbf{P}$ by the following steps:

1. In the equations of transfer, put $(n + 1)^{\text{st}}$ iterates on the integral side and n^{th} iterates on the differential side.
2. Whenever time derivatives occur, eliminate them by using the field equations of continuum mechanics, on the assumption that these latter are satisfied at each stage of iteration.

For homothermal simple shearing, the iterative system is obtained at once from $(10.14)_{1,2,3}$:

$$\begin{aligned} -\pi_{xy}{}^{\{n+1\}} &= \mathfrak{T}\pi_{yy}{}^{\{n\}} - \tfrac{2}{3}\mathfrak{T}(\pi_{xy}{}^{\{n\}})^2, \\ \pi_{yy}{}^{\{n+1\}} &= 1 + \tfrac{2}{3}\mathfrak{T}\pi_{xy}{}^{\{n\}}\pi_{yy}{}^{\{n\}}, \end{aligned} \tag{10.29}$$

where

$$\boldsymbol{\pi} \equiv \frac{\mathbf{P}}{p}. \tag{10.30}$$

† See Part I of the paper by IKENBERRY & me, "On the pressures and the flux of energy in a gas according to Maxwell's kinetic theory, I," *J. Rational Mech. Anal.*, **5**:1–54 (1956).

It is easy to solve this system by recurrence for the first few values of n and also to prove a limit theorem: *The iteration converges when* $\mathfrak{T} < \frac{1}{3}\sqrt{2}$, *and*

$$\boldsymbol{\pi}^{\{n\}} \to \frac{\mathbf{P}^{\mathrm{D}}}{p^{\mathrm{D}}} \qquad \text{as } n \to \infty. \tag{10.31}$$

The iteration diverges if $\mathfrak{T} > \frac{1}{3}\sqrt{2}$. Thus when $\mathfrak{T}$ is small enough, the iterative scheme does yield, in the limit as $n \to \infty$, the correct dominant parts of the exact solution.

In the practice of the kinetic theory, however, iterative processes are not used in this way. Rather, they are carried two or three steps so as to yield approximations to the constitutive equation. These approximations are then put into the field equations of continuum mechanics (8.22) so as to get partial differential equations, as is done in continuum mechanics itself. Now the above result shows that the approximations *to the constitutive equation* improve as n increases, but they do not give us any information about the results of solving an *initial-value problem* with those approximations. This matter, too, we can check, since we have the exact solutions (10.18). We have shown that these are approximated, as $t \to \infty$, by the dominant parts $\mathbf{P}^{\mathrm{D}}$. The ratios of these dominant parts $\mathbf{P}^{\mathrm{D}}/p^{\mathrm{D}}$ are approached, in turn, by the Chapman-Enskog iterates $\boldsymbol{\pi}^{\{n\}}$ as $n \to \infty$. The question now is, by taking the iterates for a large enough but fixed n, putting them into the field equations, and solving the initial-value problem, do we get results which in time approximate the dominant parts $\mathbf{P}^{\mathrm{D}}$? The problem may be viewed as one of inverting the order of limits, or of justifying the neglect of small terms in differential equations.

Here the reply is negative. If we write ${}^{[n]}\mathbf{P}$ for the result of the integration, we can show that

$$\frac{{}^{[n]}P_{km}}{P_{km}{}^{\mathrm{D}}} \to \text{either } 0 \text{ or } \infty \qquad \text{as } t \to \infty, \tag{10.32}$$

and the approach is exponential. That is, *for no finite n* do the pressures as functions of time gotten by *applying* the result of the Chapman-Enskog process settle as $t \to \infty$ into good approximations to the true ones; quite the reverse, they show an entirely wrong trend in time and in the end are either infinitely larger or infinitely smaller than the correct pressures, whatever be the initial conditions.

In this particular flow, as in gross rest, we may determine also the higher moments. The third moments satisfy an ordinary differential system of tenth order, which can be solved. Since the temperature is uniform at each time, we might expect all third moments to relax to zero, but this is not always true. In fact, it is possible to show that three of the third moments, while they relax if $\mathfrak{T} < 3/\sqrt{2}$, but more and more slowly the larger is $\mathfrak{T}$, increase exponentially with time if $\mathfrak{T} > 3/\sqrt{2}$. In rough terms, the class of flows considered is *unstable in energy* if $\mathfrak{T}$ is sufficiently large.

The results presented in this lecture are meager, but they are proved. They are the only exact theorems about viscosity yet derived from the kinetic theory. Much more, indeed, may be found in the literature, and much of that found may be true, although it rests so far upon purely formal systems of approximation or other plausible rather than precise arguments. The example of homothermal shearing shows how delicate may be the position of such formal processes. It pertains just as much to continuum mechanics as to the kinetic theory, for it illustrates the fact, obvious enough, though, I fear, but dimly seen in "applied" circles, that a constitutive equation which is as it stands a good approximation may lead to wrong results when it is used to solve a problem based on differential equations.† The reason is, of course, that a small function does not necessarily have a small derivative. In theories so delicate as MAXWELL's kinetic theory and modern continuum mechanics, matters of this kind cannot be dismissed by rituals or tergiversation.

In this lecture and the two preceding on the kinetic theory, I have emphasized the differences and the similarities between its predictions and those of classical hydrodynamics. At the same time, I have tried to make clear the entire consistency of results, despite a thoroughgoing difference of approach and method, with those of modern general continuum mechanics. The only domain of doubt centers upon the irreversibility principle. While we have not found the form of the total caloric H for the particular solution considered here, we know that it satisfies BOLTZMANN's H-theorem in the general form (8.41), which follows from the Maxwell-Boltzmann equation for all solutions. Does H satisfy also the Clausius-Duhem inequality of continuum mechanics?

To approach this problem, we shall use η_{M} and η_{K} in the senses defined by (8.36) and (8.38), respectively. That is, η_{M} is the caloric for the Maxwellian molecular density defined by the parameters ρ, θ, and $\dot{\mathbf{x}}$ and hence given explicitly by (8.36), while η_{K} is defined by (8.38) from the molecular density governing the problem at hand. By (8.8), $\dot{\varepsilon} = \theta\dot{\eta}_{\mathrm{M}}$ here, and by (2.26), then, the internal dissipation δ is given by

$$\rho\delta = \rho\theta(\eta_{\mathrm{K}} - \eta_{\mathrm{M}})^{\cdot} + w, \tag{10.33}$$

where w is the stress power. Since $\boldsymbol{\gamma} = \mathbf{0}$ in this problem, the Clausius-Duhem inequality (2.47) reduces to Planck's inequality (2.27), so the

† *Cf.*, for example, the instability of solutions for the fluid of second grade, as explained in *The Non-Linear Field Theories of Mechanics*, *op. cit.*, Section 123.

question is:

$$\rho\theta(\eta_{\text{K}} - \eta_{\text{M}})^{\cdot} + w \geqq 0? \tag{10.34}$$

I cannot answer this question, but I can exhibit solutions in which w is negative for a period of time†. Indeed, by (10.11), all we have to do is find a solution in which P_{xy} has the same sign as K. This means that the shear stress $-P_{xy}$, instead of pointing in the direction of flow, points against it (Figure 13). Traditional views in mechanics would regard such a result as outlandish: We must push the plates in the direction *opposite* to the gas flow within, and in so doing, we shall extract work from the gas rather than doing work on it! However, this queer effect is not inconsistent with the kinetic theory. Indeed, MAXWELL's relaxation theorem (9.3)

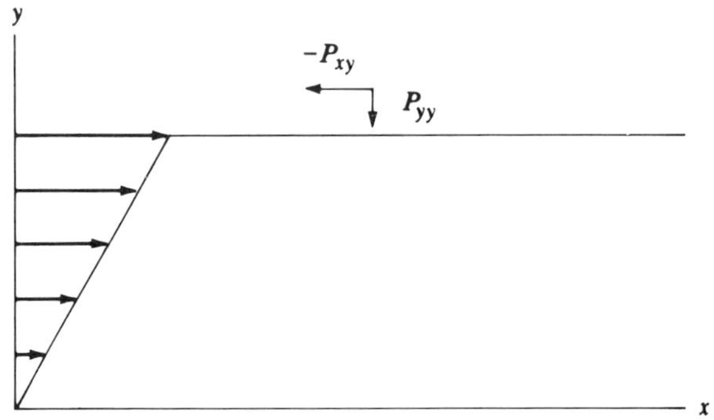

FIGURE 13. Shear stress pointing upstream, a possibility according to the kinetic theory.

comes close to it. There, we may give $\mathbf{P}(0) - p(\rho, \theta)\mathbf{1}$ any value we like, provided only it be consistent with (8.9), and each component of $\mathbf{P} - p\mathbf{1}$ will maintain its initial sign forever. Shear stress pointing upstream in shearing is scarcely more repugnant to elementary mechanics than is any shear stress at all in a fluid at rest, and both may be explained on the same grounds. The case when $K \neq 0$ is a little different from gross rest, however, in that the transient parts of the solution (10.18)–(10.20) are oscillatory. The constant A plays the critical role. From (10.18) we see at once that the condition

$$A > 0 \tag{10.35}$$

† In presenting this result in 1956, I described it as showing that "the H-theorem does not always imply the entropy principle." See Section 50 of my paper, cited above. At that time the general theory of entropy production was obscure, and I did not see that the Clausius-Duhem inequality need not imply non-negative stress power in all isochoric motions. *Cf.* Lecture 2.

is *necessary* in order that the pressure be positive, as it must be for all solutions in the kinetic theory. Moreover, (10.35) is *sufficient* that the dominant solution $\mathbf{P}^{\mathrm{D}}$ yield positive stress power for all time, but it is *not sufficient* that the full solution $\mathbf{P}$ have positive stress power initially and for a short time thereafter. Indeed, one can find initial conditions† such that for the $\mathbf{P}$ corresponding to them, $\mathbf{n} \cdot \mathbf{P}\mathbf{n} > 0$ for all $\mathbf{n}$ and all time but $-KP_{xy} < 0$ in the time interval $[0, t_0]$ for a certain positive t_0. In a solution of this kind, a finite volume of gas at first gives out gross work for a while. Of course, finally this work is used up, since $-KP_{xy} \to +\infty$ for *every* non-trivial solution as $t \to \infty$.

As suggested by MAXWELL's relaxation theorem, we cannot expect that pressures and fluxes of energy corresponding to *all* solutions of the Maxwell-Boltzmann equation be of the kind elementary mechanics or even modern continuum mechanics teaches us to regard as natural. Rather, for interpretation it is the dominant solutions we should study: We should try to prove their existence and determine their properties. In the problem of homothermal shearing, by $(10.22)_2$ and (10.13) the dominant stress power w^{D} has the simple form

$$w^{\mathrm{D}} = \frac{A\mathfrak{r}}{\frac{2}{3}\tau} e^{\mathfrak{r}t/\tau}. \tag{10.36}$$

We do not know for certain whether a dominant caloric $\eta_{\mathrm{K}}{}^{\mathrm{D}}$ exists, but there is good reason that it should‡. If it does, by (10.34) and (10.36) we should regard the following as the *Clausius-Duhem inequality for the kinetic gas* in homothermal shearing:

$$\rho\theta(\eta_{\mathrm{K}}{}^{\mathrm{D}} - \eta_{\mathrm{M}})^{\cdot} \geqq -\frac{A\mathfrak{r}}{\frac{2}{3}\tau}. \tag{10.37}$$

By (10.35), a sufficient condition for it, for all $\mathfrak{r}$ and τ, is

$$\dot{\eta}_{\mathrm{K}}{}^{\mathrm{D}} \geqq \dot{\eta}_{\mathrm{M}}. \tag{10.38}$$

That is, if the dominant kinetic caloric $\eta_{\mathrm{K}}{}^{\mathrm{D}}$ increases at least as fast as does the caloric corresponding to equilibrium at the present density,

† Equations (50.12) in my paper, where proofs are given for all statements made in the text above.

‡ Specifically, I proved that the dominant parts of the second moments exist and are in fact exact solutions of equations of second moments corresponding to certain particular initial conditions. From formulae given in my paper, a corresponding theorem can be proved for the third moments, and I can see no difficulty in principle for the moments of any order. The expectation of the logarithm is another matter. We should have to prove first that the particular initial conditions, infinite in number, which exactly annul the damped parts of all moments yield a caloric $\eta_{\mathrm{K}}{}^{\mathrm{D}}$, and second that this $\eta_{\mathrm{K}}{}^{\mathrm{D}}$ is the asymptotic form for the calorics η_{K} corresponding to all other possible initial conditions for homothermal shearing.

temperature, and velocity, then the Clausius-Duhem inequality is satisfied by the dominant solution. Of course, $\dot{\eta}_M \to \infty$ exponentially as $t \to \infty$. Since we do not know how to estimate $\dot{\eta}_K{}^D$, the conditions (10.37) and (10.38) are not effective, but they are suggestive.

At present, only two axioms of irreversibility that are both explicit and sufficiently general to cover anything beyond merely linear situations have been put forward: the Clausius-Duhem inequality of continuum thermodynamics and the Maxwell-Boltzmann collision axiom in the kinetic theory of gases. By its nature, the latter, interpreted by means of its corollary, the H-theorem, cannot include the former as a special case. The kinetic theory has not yet been developed mathematically to the point where we can determine, conversely, whether the former includes the latter, or whether the two are always consistent in the cases when both might apply.

INDEXES

NAME INDEX

SUBJECT INDEX

This book was set in Modern by The Maple Press Company, and printed on permanent paper and bound by The Maple Press Company. The designer was Marsha Cohen; the drawings were done by Harry Lazarus. The editors were Donald K. Prentiss and Laura Warner. Stuart Levine supervised the production.